Partisans, Guerillas, and Irregulars

Partisans, Guerillas, and Irregulars

Historical Archaeology of Asymmetric Warfare

Edited by Steven D. Smith and Clarence R. Geier

The University of Alabama Press
Tuscaloosa

The University of Alabama Press
Tuscaloosa, Alabama 35487-0380
uapress.ua.edu

Typeface: Minion

Cover image: Don Troiani, *Huck's Defeat, Battle of Williamson's Plantation*, 2014, oil on canvas; private collection/Bridgeman Images
Cover design: David Nees

Library of Congress Cataloging-in-Publication Data

Names: Smith, Steven D., editor. | Geier, Clarence R., editor.
Title: Partisans, guerillas, and irregulars : historical archaeology of asymmetric warfare / edited by Steven D. Smith and Clarence R. Geier.
Description: Tuscaloosa, AL : The University of Alabama Press, [2019] | Includes bibliographical references and index.
Identifiers: LCCN 2018049291| ISBN 9780817320201 (cloth) | ISBN 9780817392345 (ebook)
Subjects: LCSH: Military archaeology. | Asymmetric warfare. | Guerrilla warfare. | Irregular warfare. | Battles.
Classification: LCC CC77.M55 P37 2019 | DDC 355—dc23
LC record available at https://lccn.loc.gov/2018049291

To Michael C. Scoggins, scholar, musician, and friend
1953–2019

Contents

Illustrations

FIGURES

TABLES

Preface

The impetus for this volume came from the 2014 Fields of Conflict conference in Columbia, South Carolina. Inaugurated in 2000, this international biennial hosts conflict archaeologists devoted to, and specializing in, the study of battlefields, camps, forts, and other military sites. At this particular conference, several papers were not focused on the classic, large, conventional, set-piece battles that were common in past conferences. Instead, these papers discussed examples of raids, ambushes, and other small-scale skirmishes that, on the surface, one would imagine did not leave enough of an "archaeological signature" to be approachable through archaeology. Yet using document research, fine-grained landscape analysis, modern remote sensing techniques, and tenacity, archaeologists were revealing, rediscovering, and often reinterpreting these brief events in history. Another common theme was that these *petite guerre* encounters took place between groups of significantly different cultural and military traditions and capabilities, most taking on a "David vs. Goliath" character, shaping the definition of asymmetric warfare. Working with those Fields of Conflict papers and additional archaeological studies of asymmetric conflict, we have transformed some of the best into chapters for this volume. We believe this is the first such volume and hope it will lead to many others.

Finally, we would like to thank all those who worked hard to improve our manuscript, including our anonymous reviewers and the editorial staff of the University of Alabama Press. We especially thank Wendi Schnaufer, senior acquisitions editor; our copyeditor, Jessica Hinds-Bond; and Kelly Finefrock-Creed, project editor.

Steven D. Smith
Clarence R. Geier

Partisans, Guerillas, and Irregulars

Introduction

An Archaeology of Asymmetric Warfare

Steven D. Smith

Defining Asymmetric Warfare

This book examines archaeological contributions to the study of asymmetric warfare. What is asymmetric warfare? Apparently there is not a clear consensus among scholars. A 2007 book entitled *Asymmetric Warfare* by Rod Thornton had an entire section called "A definition of asymmetric warfare," and in the end, Thornton gave up trying to define it precisely, noting, "We do not really need to define asymmetric warfare as much as to understand what it means" (Thornton 2007:21). Nevertheless, Thornton's book includes various types of conflict that have common characteristics: the weak against the strong, nonconventional against conventional combatants, and tactics ranging from ambush to the use of box cutters and computer hacking.

The current US Army field manual, *Army Special Operations Forces: Unconventional Warfare*, includes asymmetric warfare within the genre of irregular and unrestricted warfare that uses "classic techniques of guerilla warfare, terrorism, sabotage, subversion, and insurgency" and defines unconventional warfare as "operations conducted by, with, or through, irregular forces in support of a resistance movement, an insurgency, or conventional military operations" (United States Army 2008:1-1, 1-2). Major David L. Buffaloe, a congressional liaison with the Joint Improvised Explosive Device Defeat Organization, defines asymmetric warfare more broadly as "population-centric nontraditional warfare waged between a militarily superior power and one or more inferior powers which encompasses all the following aspects: evaluating and defeating asymmetric threat, conducting asymmetric operations, understanding cultural asymmetry and evaluating asymmetric cost" (Buffaloe 2006:17). Still another, even broader definition of asymmetric warfare—under the term *strategic asymmetry*—is provided by the Strategic Studies Institute, which defines it as "acting, organizing, and thinking differently than opponents in order to maximize one's own advantages to exploit an opponent's weaknesses, attain the initiative, or gain greater freedom of action" (Metz and Johnson 2001:5). Today the term has come to describe any warfare where the opposing combatants are at opposite ends of a political or tactical spectrum, of unequal abilities or means to engage on equal footing. In other words, it is anything that is not considered "conventional" warfare. In the end, the contributors in this book turn to *Merriam-Webster*, which defines asymmetric warfare as conflict "between opposing forces which differ greatly in military power and that typically involves the use of nonconventional weapons and tactics." Critical to our definition is the idea that asymmetric warfare is warfare between combatants who are unequal in military power, politics, population, or technology. This definition incorporates all definitions noted to describe closely related types of nonconventional, usually mobile, warfare conducted by lightly armed combatants throughout the long history of human conflict. Tactically, this type of warfare consists typically of raids, ambushes, surprise, harassment, skirmishing, and terrorism. As such, the tactical roots of asymmetric warfare go deep into prehistory and primitive warfare and the very origins of human warfare (Gat 1999, 2000a, 2000b; Keeley 1996; Otterbein 2004; Turney-High 1971).

Indeed, asymmetric warfare as defined in this book is an umbrella term encompassing a long history of warfare under other names: guerilla warfare, irregular warfare, partisan warfare, low-intensity warfare, and small wars (*la petite guerre*). Asymmetric warfare is simply the latest term in use for such warfare by academics and the US Department of Defense (Barnett 2003; Babits 2004).

The tactics used in guerilla, irregular, and partisan warfare are identical, and in general usage the terms are often synonymous. It is useful, however, to note their subtle distinctions that may increase our opportunities for a more nuanced understanding of asymmetric warfare beyond mere tactics. For instance, guerilla warfare, still the most widely used term, originated during the Spanish partisan resistance to Napoleon's invasion of the Iberian Peninsula (Dederer 1983:32; Laqueur 1976:7). Guerrilla warfare is "a type of warfare characterized by irregular forces fighting small-scale, limited actions, generally in conjunction with a larger political-military strategy, against orthodox military forces" (Asprey 1975:xi). Mao Tse-tung is recognized as the modern master of guerilla warfare. As he saw it, guerilla warfare is a strategic defensive posture, a stage in the development of full-scale war that begins with partisan warfare and evolves to mobile warfare, then to guerilla warfare, and eventually to regular (conventional) warfare where standing conventional armies maintain the offense (Weigley 1970:16). Mao, applying guerilla tactics, noted that "the enemy advances, we retreat; the enemy camps, we harass; the enemy tires, we attack; the enemy retreats, we pursue" (Mao in Dederer 1983:14–15). Mao saw guerillas as professional soldiery engaged in these tactics. They are a step above partisans, who are civilian-soldiers engaged in the use of those same tactics.

The term *partisan* has earlier roots, emerging during the sixteenth century to denote a guard's officer and later lightly armed cavalry. By the American Revolution the term had broadened to include any light mobile soldiery and their tactics and included citizen soldiers (Stevenson in Hoyt 1811:78; Weigley 1970; Wright 1931:193). It was also applied to mobile light infantry and cavalry, such as Lieutenant Colonel Henry Lee's Partisan Corps, the eighteenth-century equivalent of a combined arms unit consisting of cavalry and infantry (Lee 1998 [1812]). George Washington preferred the term *Petit Guerre* (*sic*) (Dederer 1983:32). The word *partisan* today has further evolved to denote any member of an irregular force consisting of a local population resisting a foreign occupation, such as the partisans of Yugoslavia during World War II (Heilbrunn 1962).

Irregular warfare is another general term under the umbrella of asymmetric warfare, interchangeable with *partisan* and used to describe a broad range of nonconventional tactics and warfare of weak against strong (Weller 1977). In use, it is often attached to Native American warfare and militias fighting in North American forests during the seventeenth and eighteenth centuries (Ferguson 1979; Waghelstein 1995). Contemporaries called it the "skulking way of war" (Malone 1991).

By definition, asymmetric warfare must have originated with the rise of complex societies where the potential for unequal adversaries became possible. It is difficult to conceive of unequal hunter-gatherer bands in early prehistory engaging in protracted warfare, although the tactics of raid, ambush, and skirmish unarguably originated as hunter-gatherer warfare. Admittedly, archaeological evidence of prehistoric asymmetric hunter-gather warfare would be nearly impossible to find. Therefore, we see the origins of asymmetric warfare as associated with the origins of the state, where complex societies came in conflict with their lesser neighbors.

We can see the philosophical origins of asymmetric warfare in the classic treatise *The Art of War* by Chinese strategists writing under the name Sun Tzu in the fourth century BC, although they were not necessarily practitioners of asymmetric warfare. The Sun Tzu took an indirect approach to warfare, emphasizing deception, rapid maneuver, attacks on an enemy's weaknesses, and the avoiding of its strengths. These are classic guerilla tactics that influenced Mao Tse-tung and countless others (Joes 1996:231). Beyond the Chinese, the Scythians' defensive warfare against the Persians under Darius I around 512 BC and of Judas Maccabeus against the Syrians around

165 BC are early examples of asymmetric warfare. The Romans too would meet guerillas in the mountains of Spain (147–139 BC) as would Napoleon, but the Romans proved more ruthless in counterinsurgency and eventually conquered the guerillas (Joes 1996:3–4).

Another example is the Greek Dacian insurgency against the Romans around 106 AD. Archaeologists Jeremy Miller, Álvaro Ibarra, and M. Scott Harris have used a multidisciplinary analysis of the Dacian landscape to demonstrate the effects of the insurgency on the Roman occupation and counterinsurgency. As with modern attempts at counterinsurgency, the Romans placed their *castra* (fortified camps) not only for strategic military advantage but also to symbolically dominate the political and social landscape. Meanwhile, a prolonged insurgency made the cost of Roman occupation too high, and the Romans withdrew as their global power declined. This is food for thought for our modern times (Miller et al. 2014).

Indigenous peoples across the globe were and still are expert practitioners of the art of irregular warfare. In North America, from the sixteenth-century conquistadors to the last tribal remnants in the nineteenth-century American West, Native Americans repeatedly won battles and lost wars.

Historical Archaeology and Asymmetric Warfare

This volume presents ten case studies of low-intensity conflict from the sixteenth to the twentieth century through the perspective of historical archaeology. It includes examples of irregular, guerilla, and partisan warfare, which we incorporate under the umbrella of asymmetric warfare. Whereas today the world's great superpowers still threaten each other with nuclear and conventional wars, the world's lesser nation-states, and nonstate actors, continue practicing irregular, partisan, or guerilla warfare as a means to achieve political, military, and ideological goals. These ancient, indeed primitive, forms of warfare have proven to be the most feasible means for weak social groups to achieve victory over both equal and stronger foes. Likewise, the strong have used the same methods and tactics to subdue the weak in counterinsurgency operations throughout history. The studies herein cover events in American and world history: the French and Indian War, the American Revolution, Indian Wars in the Southeast and Southwest, the Civil War, Reconstruction, Prohibition, and World War II. Most of our examples occurred on the North American continent; all provide lessons globally applicable.

Most of the studies in this volume are mere blips in the protracted and global history of human conflict. Yet they were singular events within a much larger and more complex historic movement. Whether they succeeded and contributed to historic movements or failed and were lost in time, such moments are basic to virtually every social, political, economic, and military movement that has taken place in the historic era. For this reason alone, they are important to study and understand.

As the import of these events needs to be understood, the lack, absence, or questionable validity of historical accounts becomes a handicap. Although retained in local memory or tradition, and despite their potential importance, they are poorly and incompletely addressed in the historical record, containing no specific details of the "who, what, when, where" of the event. In response, the highly multidisciplinary field of historical archaeology, which seeks to locate and interpret remaining documentary and physical, material evidence of past events, has become established. Within this larger field the internationally supported focus of conflict archaeology has been involved with the interpretation of military sites and events from antiquity through the Vietnam War dealing not only with understanding a particular military action but its immediate and long-term human and cultural impact. By using examples, primarily from American history, the contributors introduce the reader to the growing study of asymmetric warfare and to the historic importance of their topic. At the same time, they illustrate how the wide range of traditional and new methods and techniques of historiography and archaeology can be applied to expose critical actions, sacrifices, and accomplishments of competing groups representing opposing philosophies and ways of life that are otherwise lost in time.

Chapter Overviews

Charles M. Haecker (chapter 8) provides a four-century overview of these "dirty little wars" beginning with Cortes's 1519 expedition. Haecker's archaeological analysis demonstrates how the strong underestimated the weak, using the example of the Spanish and, later, Americans attempting to subdue Native Americans, particularly the Apache. The Spanish and Americans were often not only outmaneuvered but, surprisingly, also outgunned on the battlefield. The highly mobile Apache used the rough western landscape to their utmost advantage and adapted their tactics to the immediate situation. They quickly adopted and adapted the advanced military technology brought by the Spanish and Americans to their region. Meanwhile, larger, complex societies like those of the Spanish and Americans emphasized standard tactics and weaponry and were seemingly reluctant to adapt.

Perhaps this reluctance to adapt was the inevitable product of complex society. Michelle Sivilich's study (chapter 5) of formalized training at the US Military Academy, West Point, provides a further example of the failure of the US military to adapt to indigenous warfare. Her study focuses on another harsh landscape, that of the Florida swamps. During the Seminole Wars (1817–1818, 1835–1842, 1855–1858), the American officer corps attempted to apply its engineering training toward fighting the Seminoles in a hostile environment unsuited to such conventional warfare. This training was fashioned after eighteenth-century European models in which soldiers were prepared to fight conventional wars that incorporated heavy fortification, battle lines, and mass attacks, typical of the Napoleonic era. Such tactics were applicable to the European landscape but not to the swamps and marshes of central Florida. When the need arose, young officers were sent into battle with little or no experience with irregular warfare.

Both chapters (Haecker and Sivilich) provide a window into understanding how strong nation-states are often defeated on the battlefield by militarily weaker forces. The American military failures against the Native Americans in the American West were largely due to an inability to adapt to Native American tactics and mobility, combined with inferior arms. Likewise, conventional warfare tactics taught by the US military in antebellum America failed its soldiery in the swamps of Florida. These examples of the inability of the American military to adapt to irregular warfare beg the question as to why the lessons of the American Revolution and earlier warfare were not better applied. By the time the US military entered Florida to remove the Seminoles, nearly 50 years had passed since the American Revolution. This was a revolution in which irregular and partisan warfare played a significant part in defeating conventional British forces and equal foes in the form of Loyalists and Native Americans. The only explanation, or perhaps the easiest explanation, is that the American military leadership at the highest levels failed the soldiers on the ground. Poor policy decisions at the highest level seem to be the common error across the centuries.

As noted, on the American colonial frontier the militia had better success in counterinsurgency. W. Stephen McBride and Kim A. McBride (chapter 1) examine forting up along the colonial frontier of eighteenth-century western Virginia. The settlers' strategy was to construct blockhouses and palisade their homesteads to resist Indian raids. Local militia scouts roamed the wilderness keeping an eye out for raiding parties. When Indian forays were spotted, the settlers would rush to the forts for protection. The Indians still managed to surprise and ambush settlers outside of the forts, but rarely could they maintain a protracted siege against a strong fortress. The settlers' defensive strategy of forting up was hardly a European or Euro-American invention. Of course Europeans had a long history of palisaded villages going back at least to the Iron Age; however, the settlers probably picked up the strategy from Native Americans. Archaeologists have noted that Native Americans had been palisading their villages since prehistory

(Dye 2006; Keeley 1996:55–58, 190–193, Table 3.2; Malone 1991). Some scholars note that the Indians abandoned such fortifications with the introduction of total war by Europeans. Prolonged siege warfare and fire could turn a palisaded village into a death trap (Lee 2004:767). Meanwhile, by the late eighteenth century Native Americans were reduced to seasonal forays and opportunistic raids running up against Euro-Americans who had learned to build structures strong enough for temporary defense.

Michael C. Scoggins and Steven D. Smith (chapter 3) demonstrate that backcountry militias could be quite successful on the offense. The dawn attack on Loyalists camped around James Williamson's plantation shortly after the fall of Charleston, South Carolina, was a classic raid of complete surprise. The Loyalists were defeated before they understood the threat. The archaeology at the battle site indicates that surprise and overwhelming firepower gained the day. Clearly, by the time of the American Revolution, colonists had the tactics and methods of asymmetric warfare well in hand.

Irregular warfare has also been used to describe the violent actions of banditti, bushwhackers, and feuding parties. Douglas D. Scott (chapter 6), Carl G. Drexler (chapter 7), and Kim A. McBride (chapter 9) all examine these forms of low-intensity conflict in which combatants are operating beyond the bounds of legal authority. Scott examines an 1856 clash between pro- and antislavery forces in Bleeding Kansas, before the Civil War. Although Scott provides a detailed overview of the battle, his focus is on the weaponry and landscape of battle revealed by archaeological study. Scott's study reinforces a pattern seen by Haecker, by Adrian Mandzy (chapter 10), and to an extent by Steven D. Smith (chapter 4) in partisans at Snow's Island, South Carolina. That pattern is the variety of weaponry used by irregular forces. Indeed, as Scott asserts, the variety of ammunition types and calibers is probably an archaeological signature for asymmetric warfare.

Carl G. Drexler looks at banditti and bushwhackers on the Confederate home front in Arkansas (chapter 7). It is not widely appreciated that resistance to the Confederacy was widespread throughout the South during the war. Further, there were those in the largely rural South who took advantage of the chaos of war to settle old slights or feuds. Still further, disillusioned Confederate soldiers left the ranks and returned home to discover that all was not well in their part of the country and, in response, formed armed groups for defense. Ultimately these groups clashed at various times with Confederate and Union conventional forces. It was difficult for conventional forces to sort through such groups, but the US Army did attempt to classify them so as to provide some rules of engagement in what is now known as the Lieber Code. Interestingly, Drexler finds that the Confederacy used a timeworn but effective method of controlling such groups by simply controlling the landscape. A well-placed fortification at a critical choke point along a well-traveled route checked or at least controlled movement. Like Steven D. Smith in his investigation of Francis Marion's partisans (chapter 4), Drexler notes that forts and redoubts are not limited to conventional warfare but play a significant role in irregular warfare also.

Kim A. McBride (chapter 9) balances this volume with a study of post–Civil War kin-based feuding, demonstrating how such unconventional conflict can linger beyond a conventional war's end. Feuding may be considered at the low end of conflict intensity, just a grade above homicide or gang violence, which involves a series of revenge killings (Keeley 1996:29; Otterbein 2009:44–45), yet it does involve group-on-group sustained violence. The Hatfield-McCoy feud is one of the most famous, and the Hatfield raid against the Randall McCoy homestead was battle in itself. McBride asserts that the level of violence in the raid ultimately brought the feud to a close. Further reflection on this type of low-intensity conflict is the subject of Clarence R. Geier's concluding thoughts.

Three chapters in this volume address partisan warfare, a type of asymmetric warfare with an evolving etymology. In chapter 2, Wade P. Catts archaeologically examines an action at Cooch's Bridge in 1777 during the American Revolution. Here partisan warfare is used in its original meaning: light troops used for scouting, skirmishing, and screening in support of the larger conventional forces. The chapter's significance for this volume lies in part because one of the battle's participants was Johann Ewald, who returned to Europe after the war and not only wrote of his experiences but also drafted a treatise, or military manual, on partisan warfare (Ewald in Selig and Skaggs 1991). *Abhandlung über den Kleinen Kreig* (*Treatise on Partisan Warfare*) provided a practical approach or how-to for the use of light troops. As the editors of Ewald's treatise note, light troop use had a long history before Ewald, so the book's impact (indeed that of the American Revolution as a whole) on European tactics is debatable, but his axioms are still applicable to small wars today (Selig and Skaggs 1991:29).

Chapter 4 by Steven D. Smith examines partisan warfare in a different context and meaning. By the American Revolution, the term *partisan* included not only light troops but also those who used light troop tactics, including civilians. Although Francis Marion, widely recognized as South Carolina's ultimate partisan warrior, was a lieutenant colonel and later brigadier general of conventional forces, he earned his fame as the "Swamp Fox" as a partisan guerilla. Marion was labeled a partisan by both his British enemies and his biographers (Simms 1845; Griffith 1978:7). He was ordered by his Continental army commander, General Nathanael Greene, to conduct a "partisan war" (Greene 1991 [1780]:520). Thus Marion was both a partisan and someone in command of partisans. To further confuse the issue, partisans at that time were often synonymous with militia and were also called militia both in contemporary writing and in military history. This chapter examines the community of partisans who made up Marion's forces from August 1780 until the end of the war. Smith concurs with Mao Tse-tung, who would later write that partisans need a secure base and the support of the local population to survive. Surprisingly, there is a great deal of archaeological evidence of Marion's partisans around his famous hideout on Snow's Island, South Carolina.

Adrian Mandzy's contribution (chapter 10) examines a partisan ambush during World War II. These partisans were members of the Ukrainian Partisan Army, formed to resist both German and Soviet intrusion into their country. Largely unsupported, they resisted mainly by the capture of enemy weapons, which was reflected in the variety of spent ammunition recovered by archaeologists at the ambush site. Mandzy's archaeological investigation reveals that the ambush's success was due to adherence to tactical manuals, well-placed automatic weapons, and an equally well-placed individual rifleman with a clear field of fire. Obviously, the tactics of asymmetric warfare can be learned and understood by conventional forces if they adapt to the situation on hand.

KOCOA

Many of the chapters use KOCOA as an analytical tool in the study of battlefields. KOCOA is an acronym for Key Terrain, Observation and Fields of Fire, Cover and Concealment, Obstacles, and Avenues of Approach. It is a systematic method of looking at the battlefield landscape from the perspective of the soldier on the ground attempting to achieve a military objective. The US military version of this method is OCKOA. This form of battlefield landscape analysis traces back to military historian Lieutenant Colonel A. H. Burne, who recognized in the 1950s that by walking the battlefield, he could determine the "solution of an obscurity by an estimate of what a trained soldier would have done in the circumstances," which he called "Inherent Military Probability" (Keegan 1977:33–34). Key Terrain is any landscape feature necessary to gain an advantage over the enemy. Observation and Fields of Fire are often the same features as Key Terrain, allowing observation of the enemy and clear lanes for firing against an enemy. Cover and Concealment are landscape features that cover or conceal maneuvers from the opponent. Obstacles are natural landscape or human-constructed features that block or restrict movement, and Avenues of Approach are the likely features (roads) that allow or accelerate movement to and from the battlefield (Babits

2014:265–267). Obviously, the importance of certain features changes based on the military technology of the time. The top of a hill might be critical in eighteenth-century warfare but dangerous in World War II. The KOCOA method is extremely useful for archaeologists attempting to determine and understand the flow of battle, especially when the landscape has changed significantly since the battle occurred.

Step one in KOCOA analysis is to reveal defining features by an analysis of any and all historical documents related to the battle. Eyewitnesses describing an attack might mention crossing a ravine or climbing a hill. These landscape nodes become defining features to be discovered by geospatial digital analysis or an on-site survey of the battlefield. From there, the battle unfolds across the landscape for the archaeologist and can be confirmed through archaeological investigation, often using metal detectors. The technique can be used in reverse also. Perhaps the main battlefield is known. From there, KOCOA analysis allows the archaeologist to determine likely avenues of approach, increasing our knowledge of events leading to the main combat location.

The contributors to this book take a broad perspective on asymmetric warfare. We see it as a modern term to cover a variety of similar forms of irregular warfare where the opposing forces are not equal in societal complexity, military strength, technology, or population. In such situations, the weaker combatants usually resort to tactics that avoid stand-up, face-to-face battle. Instead, they rely on overcoming their weaknesses by mobility, raids, ambushes, and stealth. It is, as Mao would say, a strategically defensive posture, although success is the result of tactical offensive action.

We also take a broad perspective of archaeology. As applied in this volume, archaeology includes the examination of any and all forms of evidence of the past that can enhance our battlefield analysis and ultimate understanding of conflict. Material evidence of battle, landscape analysis, KOCOA, the historical and oral record, and specialized scientific analyses are all partners in this endeavor.

1
Border Warfare in Revolutionary Era West Virginia

W. Stephen McBride and Kim A. McBride

The colonization of present-day West Virginia was fraught with conflict between Native Americans and Euro-American settlers and their enslaved African Americans for much of the middle to late eighteenth century. At the heart of this conflict was the contested ownership of land, some of which transpired as part of broader geopolitical contests, first between the nation-states of England and France (French and Indian War) and then England and the United States (American Revolution). As such, this study provides an example of asymmetric warfare between established Native American tribes and Euro-American settlers in the Greenbrier Valley, West Virginia, that began in Lord Dunmore's War of 1774, which pitted the Native Americans against the British-governed settlers of western Virginia, and continued into the American Revolutionary War (1775–1783), when the British encouraged and aided Native American raids on frontier settlements. During both wars, Indians living north of the Ohio River but who laid claim to present-day West Virginia, including Shawnee, Wyandot, and western Iroquois (Mingo), launched raids into the Greenbrier Valley.

Warfare on the frontier during this time was irregular; it did not involve professional standing armies facing each other on open ground. It involved local militia and often untrained settlers, of all ages and genders, and skilled Native Americans fighting each other in the forests, in villages, in farm fields, in houses, and in forts. The settlers, as an extension of a larger nation-state, first Great Britain and later the United States, were associated with a more powerful, more populous, and more technologically advanced society than were the Native Americans. Because the settlers were living in rather isolated conditions on the frontier, however, these advantages were somewhat lessened, although they did maintain a significant population advantage over the tribal groups. Even though the Native Americans were not part of a nation-state, they often formed broad, if temporary, alliances and had the support, and supplies, including guns and ammunition, of European allies, including France in the middle eighteenth century and Great Britain in the late eighteenth century. In each case their men, while of limited numbers, were trained warriors, skilled in the arts of frontier warfare. In fact, historian Richard White has characterized this frontier warfare as a "contest of villagers," meaning that the Indians and settlers had their own "village" agenda and mode of warfare within the broader global conflicts (White 1991:367). Like most irregular or guerilla fighters, the Native Americans relied heavily on surprise and ambush, but when acting in large raiding parties they were often highly organized. In the Greenbrier Valley, in the years before and during the American Revolution, the Native Americans, seeing the growing spread of settlers as a challenge to their lands and lifestyle, were nearly always on the offensive. Accordingly, the settler militia strategy and tactics, particularly on a local level, were designed as a defensive counterinsurgency to defeat and contain the Indian raiders.

In this study, historical and archaeological investigations of skirmish and battle sites within the Greenbrier Valley, as well as broader studies and treatises on frontier warfare, offer a better understanding of the nature of the irregular frontier warfare during the revolutionary era. Particular attention is given to the strategy and tactics developed and implemented by the opposing forces. Investigations include archaeological data gathered from excavations at two forts (Arbuckle's and Donnally's) and one farmstead (James Graham's) that were at the center of Native American raids in the 1770s.

Native American Warfare

By the revolutionary era, the main Native American war parties that raided into today's West Virginia lived in Ohio and were the Shawnee, Mingo (western Iroquois), Wyandot, and, toward the end of the war, some Delaware (Calloway 2007; Downes 1968; White 1991). In 1774, the Shawnee and Mingo entered a war with Virginia, known now as Lord Dunmore's War after the Virginia governor. This war had its origin in the British-Iroquois 1768 Treaty of Fort Stanwix, in which the Iroquois ceded much of West Virginia and Kentucky to Great Britain, including land claimed by the Shawnee or Mingo. The Shawnee and Mingo disagreed with the treaty and saw the ceding of land, and the subsequent wave of European settlement, as a direct threat. Fighting broke out in the spring of 1774 after the murder of Mingo John Logan's family by settlers near the mouth of Yellow Creek (Yellow Creek Massacre) in northern West Virginia (Downes 1968:152–173).

At the beginning of the American Revolution, most of the Ohio Valley and Great Lakes tribes were politically neutral in their relations with the colonies and the British. Only the Mingo were allied with the British and were at war with the Americans, but this did not preclude individual warriors, particularly Shawnee, from accompanying the Mingo on raids (Downes 1968:189–190; Howard 1981:14). By late 1777 the Shawnee were leaning toward an alliance with the British, and the murder of the Shawnee chief Cornstalk and three companions at Fort Randolph, Point Pleasant, West Virginia, in November 1777 pushed this to reality. The Wyandot soon declared war on the Americans as well. The non-Moravian Delaware joined the British in 1781, in contrast to the Moravian Delaware who remained neutral (Downes 1968:204–210; Howard 1981:15; White 1991:389).

All these Indian groups had a long history of woodland warfare, and all had become proficient in the use of firearms, particularly the flintlock musket, well before the revolutionary era, although bows and arrows were still utilized effectively (Thwaites and Kellogg 1912:79). Period observers distinguish two types of Native raids or warfare: small-scale, *petite* or private wars that usually involved 10 to 25 warriors and did not require the approval of the confederacy, tribe, or division; and larger scale, "national" or public wars that did require this approval or declaration of war by the tribe as a whole. *Petite* warfare was generally motivated by revenge, booty (especially horses), and captives and by the opportunity for young men to gain prestige by demonstrating their courage and fighting ability (Connelly 1964; Malone 1991:7; White 1991:407).

In contrast "national" wars followed a discussion and debate at a general council of peace chiefs and war chiefs and often included questions of extratribal alliance. If war was declared, the tribe shifted to a wartime footing, and the peace chiefs surrendered control of the tribal government to the war chiefs (Howard 1981:115–116). The motives and goals of national war varied but included obtaining revenge (on a larger scale), protecting or expanding tribal lands and resources, recapturing old tribal lands and resources, controlling trade and transportation, honoring alliances, and even, though rarely, vanquishing an enemy seen as a threat (Downes 1968:10–15; Eid 1988a:150; White 1991:407). Raiding during national wars included both small (10 to 25) and large (50 to hundreds of warriors) war parties, all acting in union for broader goals.

Period observers and historians often describe the Indians' main, or even only, strategy and tactic of warfare as "surprise and ambush" (Eid 1988b:63; Mahon 1958:260; Malone 1991:21, 24; Smith 1948 [1812]; Zeisberger 1999:104), or what white observers called "the skulking way of war" (Malone 1991). Given the relatively small size of most Native American groups and the limited number of warriors available to wage war and ensure the security of the home front, this style of warfare was eminently logical. The Indians simply could not afford larger battles that could result in high casualties.

According to former Indian captive James Smith's mentor Tecaughnelango, "the art of war consists of ambushing and surprising our enemies, and preventing them from ambushing and surprising us" (Smith 1948 [1812]:41). In his treatise on Indian warfare, Smith describes how the Indians achieved this surprise and how they behaved during battle. The raiding party would leave the village, following a feast and war dance the night before, moving in single file. Small groups would often be under the command of a captain, whereas large parties would be directed by multiple captains and a war chief. Smith emphasized that the warriors were very disciplined and would quickly act on the orders of their captains, noting that "the Indians are the best disciplined troops for wooded country in the known world" (Smith 1948 [1812]:9).

Other observers of eighteenth-century Indian warfare, such as Robert Rogers, Henry Bouquet, and David Zeisberger, also commented on the warriors' discipline and ability to perform complex maneuvers in the forest (Eid 1988b:63, 68; Malone 1991:20; Zeisberger 1999:103). The raiding party used cover and concealment with scouts on the lookout for the enemy and good ambush locations. Once near the enemy, the warriors would cease hunting for food and would spread out in "scattered order" to avoid clustering and decrease the chance of being surprised and surrounded (Smith 1948 [1812]:4, 12).

Finding a good ambush location, such as a low point on a well-traveled trail or roadway route, the warriors would wait silently and patiently for days to surprise unwary travelers (Zeisberger 1999:104). They might also try to lure the enemy into an ambush by sending out a decoy or create a moving ambush by silently enveloping the enemy, often near dawn. The objective was to launch a quick attack that did not allow the target to mobilize an effective response (Eid 1988a:159). According to Smith (1948 [1812]:4, 14), just before an engagement, the Indians would move into a half-moon or horseshoe formation, with flankers on both sides of the site where the target was or was expected to appear. Once in this formation, with the enemy partially surrounded, they would fire from behind trees and quickly charge, or withdraw, as circumstances dictated. As Eid (1988a:148) stated, "late eighteenth century Indian battlefield maneuvering revealed a sophisticated use of flanking movements proceeded from a half-moon or standing position." This formation was used successfully in many large and small battles throughout the seventeenth, eighteenth, and nineteenth centuries in the east and the plains (Eid 1988a and 1988b; Otterbein 2004:211–213; Secoy 1953; Sivilich, this volume, chapter 5; Smith 1948 [1812]:5–6).

Native warriors usually did not close this half-moon, leaving the enemy an escape route that would reduce the chance of a costly fight to the death, while themselves benefitting from a surprise attack. Given their demographic disadvantage to what appeared to be a never-ending flow of white settlers, the Indians always tried to keep their casualties to a minimum and generally did not initiate an attack unless their leaders were certain of victory (Smith 1948 [1812]:12). If a battle was going poorly, Indians would usually retreat rapidly. If only one section of a battle was going badly, they would withdraw from that point but return if the pressure lessened. War parties could quickly change formations from half-moons to circles, or to hollow squares or parallel enclosing lines, as circumstances warranted (Smith 1948 [1812]:4, 12).

Smaller raiding parties, and even some larger ones, would often focus on isolated farmsteads or moving parties of militia and settlers. Using surprise and ambush, near dawn or at a time when such action would not be expected, they sought to spread panic before the Americans could adequately defend themselves (Mahon 1958:259). Most Indian warriors avoided attacking substantial fortifications, but on occasion larger war parties did make forts the target of attack or siege. Again, surprise was the initial tactic with the hope of entering the fort at a weak point or catching occupants outside of and away from their defenses. If this initial surprise attack failed, they might surround and lay siege, use fire to force defenders out, enter into negotiations to encourage surrender, or simply withdraw (Downes 1968; Kellogg 1916; Mahon 1958; Rice 1970). Whereas the capture of forts through assault, or even negotiation, was common during the French and Indian War, it was rare during the Revolutionary War, as the militia and other settlers became better organized, were often in larger numbers, and were more effective in their defense.

Another Indian strategy was the use of psychological warfare, or inflicting terror or fear on their enemies. Attacks on women and children, torture and otherwise brutal treatment of prisoners, and physical mutilation of victims created a generalized fear within the settler populations (Calloway 1997, 2007; Crawford 2008). In the case of the French and Indian War period, this fear had often led to settlers' removal to safer areas to the east. Removal of settlers from the Greenbrier Valley was less common in the Lord Dunmore's War to Revolutionary War period covered in this chapter but was still a concern of the British, then colonial, government. Fear of Indian attacks still motivated settlers to build forts and otherwise try to defend themselves.

The Euro-American Frontier Defensive System

Developed first in the French and Indian War, the Euro-American frontier defensive system was refined by the onset of the American Revolution. Composed of local militia, specialized scouts or "Indian spies," and forts, the system had the central goal of protecting established settlements and outlying farmsteads. During this period, the primary resident military force was the local county militia, modeled on English precedents. All free white men aged 18 to 50, except those with occupations such as ministers, millers, and iron masters, were required to serve. Although the governor was the overall commander, a militia was organized at the county level and was led by a county lieutenant whose staff and company officers had direct command of the men. Each county had at least one regiment that was divided into companies of approximately 20 to 80 men and officers. The county lieutenant could order the militia to service for action within the county, but to take his regiments outside of the county he had to ask for volunteers, a situation that caused much frustration among offensively minded officers (Sosin 1967; Stone 1978). Beyond their military function, militia companies could be mobilized for needed road construction and maintenance and could also facilitate tax collection. This gave great power to the militia officers within their local communities.

Local militiamen garrisoned local forts, guarded settlers, and participated in more distant military expeditions (rarely). Pension applications suggest that partial or entire companies would guard a fort from anywhere from a few days to as long as six months, moving from fort to fort as needed. Militiamen also served as security for communities made up of families occupying and working on dispersed farms and, when required, pursued Indian raiding parties. According to militiaman James Gillilan, "in the summer season [we] would all turn out in a body and work each other's places by turns—whilst some were working others would be watching and guarding—to give alarm of the approach of Indians" (Gillilan 1833–1835).

The use of "Indian spies" or scouts was another crucial part of the frontier defensive strategy. During the French and Indian War, rangers and scouts as individuals and in groups functioned offensively, gathering intelligence and attacking the enemy in their camps. By the 1770s and 1780s, spies had taken on a more defensive role, roaming over the landscape looking for enemy signs and giving warnings to a citizenry otherwise engaged in their daily lives. This was especially important in the warmer months when raiding was more of a threat. Given the widely dispersed nature of frontier farms and forts and the desire/need of most settlers to stay on their farms during the warmer months, this role was critical to frontier defense. The Revolutionary War pension application of Greenbrier Valley resident and spy Michael Swope noted that "when [spies] saw signs of Indians they would fly from Fort to Fort and give the alarm so that preparations might be made for defensive operations by the people that were Forted and that those who had ventured out to work their corn might betake themselves to the Fort before the Indians would attack them" (Swope 1833–1835).

Spies commonly worked in small groups of two to three, were generally based at a fort, and usually operated near their homes in areas where they were familiar with the terrain. Period accounts describe going out on rounds of four to eight days, with a given circuit of 48 to 113 km (30 to 70 mi). Additional spies were often posted at known passes and in advance areas during times of particular danger. Pensioner John Bradshaw reported that he "watched the gaps and low places in the mountains for thirty miles, to a point where they met the spies from Burnside's Fort" (Bradshaw 1833–1835).

The frontier forts anchored the local defensive system by providing an operational base for militia and spies and serving as places of refuge for settlers in times of danger. Like the militia, the concept of a community fort is an ancient one, stemming from long-standing European traditions. Forts could be privately built to meet individual goals such as trade or were built by the militia, often as part of a line or system of forts authorized by and paid for by the colonial government. Although descriptions of West Virginia (then Virginia) frontier forts are limited, they were generally made of logs and ranged from two-story log houses, or blockhouses, to stockades with corner bastions and internal log buildings. Some solidly built stone residences were considered "strong houses" or forts.

As towns were not well established in this region until the early 1780s, forts often served as economic, mercantile, and cultural centers as well as sources of security. Usually located in the center of a settlement cluster or neighborhood, forts were often built by, or on the land of, a prominent settler who may have donated land and materials or supervised construction to gain prestige as well as secure defense for his holding and the neighborhood. Such fort owners/builders usually held several important local and regional political or military offices during Lord Dunmore's War and the Revolutionary War period.

Each population cluster or neighborhood had one or more forts. They were always near a permanent water source such as a spring or creek, were usually on a key trail or road, and were generally on an elevated ridge or terrace that could be easily accessed by the settlers they served. Militia-built forts were more likely to be located for strategic defensive reasons, whereas private forts might not be quite as strategically placed, instead reflecting the nuances of the land and motives of their owners. The distance between forts varied with population density, areas of cleared land, and exposure to danger.

For most settlers, "forting" was highly seasonal, as described by Greenbrier Valley Revolutionary War pensioner Samuel Gwinn: "All the people of the settlements took their families to the forts in the summer months, where we farmed pretty much in common would turn out all in a body and work each other's corn and potato patches by turns whilst we would be working some would be watching for Indians" (Gwinn 1833–1835). As well, an interesting commentary on a sense of regional attachment to a fort site is provided in the memoirs of the Rev. Joseph Doddridge (1824:94–95, emphasis added): "My reader will understand by this term [fort], not only a place of defense, but the residence of a small number of families belonging to the same neighborhood. . . . The families *belonging to these forts* were so attached to their own cabins on their farms, that they seldom move into the fort in the spring until compelled by some alarm, as they called it. Thus it often happened that the whole number of families *belonging to a fort* who were in the evening at their homes were all in their little fortress before the dawn of the next morning."

Two extralocal aspects of frontier defense need to be mentioned: the establishment of state forts on the Ohio River, and government-sponsored offensive campaigns. The states of Virginia and Pennsylvania constructed and/or garrisoned larger forts on the Ohio, such as Forts Pitt, Henry, Randolph, and Nelson. These were meant to stop Indian raiding parties, to send warnings of potential threats to the settlements, and, when needed, to launch offensive campaigns into Indian country. As far as stopping Indian raids, these forts were a failure, although they sometimes succeeded as a first warning system. As bases for offensive campaigns, they were perhaps more successful. Certain of these campaigns, such as those launched by Colonel George Rogers Clark, were successful in destroying Indian villages and crops, thereby undermining their ability to wage war. At the same time, the failure of the British to constrain the colonials and protect their Indian allies undermined their prestige among the tribes. Volunteer militia as well as regular Virginia and Pennsylvania troops garrisoned these forts and participated in offensive campaigns.

Engagements from Greenbrier County

Present-day Greenbrier County, West Virginia, was the scene of a number of Native American attacks on Euro-American settler sites during the mid and late eighteenth century. Most of these raids are poorly documented. The majority were small-scale, *petite* raids that targeted isolated or vulnerable settler farmsteads or small parties of settlers or militia traveling between farmsteads or forts. Raids of this type are documented for 1755, 1763, 1774, 1776–1778, and 1780 (Kellogg 1916; Morton 1916; Rice 1970, 1986; Thwaites and Kellogg 1905, 1908, 1912). Raiding parties rarely attacked forts, but they did on a few occasions, including the 1774 and 1777 attacks on Arbuckle's Fort on Muddy Creek, and the larger scale 1778 campaign and assault on

Donnally's Fort in Rader's Valley. Two of these battle sites, Arbuckle's Fort and Donnally's Fort, plus the Graham house, the site of a reported farmstead raid, have been investigated archaeologically and are used here to illustrate frontier defense and warfare.

Petite *Warfare: Arbuckle's Fort and Graham Farm*

Arbuckle's Fort, also known as Keeny's or Muddy Creek Fort, was constructed on John Keeny's property above the confluence of Muddy and Mill Creeks of the Greenbrier River in present-day Greenbrier County, West Virginia. The fort was constructed by Captain Matthew Arbuckle's militia company in April 1774, early in Lord Dunmore's War. Although no descriptions of this fort have been located, the level of militia occupation during Dunmore's and the Revolutionary War suggests that this fort was likely stockaded and that it was the most important fort in the Muddy Creek neighborhood. Period letters and Revolutionary War pension applications suggest that the fort was garrisoned from spring to early fall between 1774 and 1782 by at least part of the Muddy Creek militia company. Captain Arbuckle initially served as militia commander, with that role later passing to Captain William Hamilton. The fort occasionally housed militiamen from more distant places such as the Shenandoah Valley (McBride and McBride 1998).

Both the 1774 and 1777 attacks on Arbuckle's Fort are poorly documented. Writing on August 1, 1774, Major James Robertson provided some insight into the first action: "this minet I got flying news of the Indians shooting at one of Arbuckle's Centery's on Mudy Creek. They say Likewise that they attacked one Kelly's yesterday about half a mile from that Fort where they Tomhak'd Kelly and Cut him Vastly, but the men from the fort heard the Noise and Ran to their Assistance and drove the Indians off before they Either kill'd or sculp'ed Kelly" (Thwaites and Kellogg 1905:103–104).

In his pension application, James Kelly confirmed that his father William died of wounds received in the 1774 raid and noted that "his father before that sent his family (Declarant among the rest) to Arbuckle's Fort" (Kelly 1833–1835). This suggests that there had been a warning before this raid. Leah Viney recorded that her family was in Arbuckle's Fort during the 1774 attack and that "she remembered seeing the burial near the fort of a man [Kelly or sentry?] who had been killed by Indians lurking about the fort" (Rodgers ca. 1900). Finally, Jacob Stevens stated "Before we came [in 1775] John Keeney's fort, on the Muddy Creek was attacked by Indians" (Stevens ca. 1840).

The second assault on the fort came on September 11, 1777, when, according to Captain John Stuart, "a number of guns were heard by sundry persons in our neighborhood supposed to be at muddy creek fort about sundown last night" (Thwaites and Kellogg 1912:80–81). In the same letter, dated September 12, Stuart asked "Capt. Hendry Smith to send a Serg.'s command at least to assist the people of muddy creek who is very few in numbers, and I am afraid will be much distress." This is all that is recorded about this attack, but earlier on September 11 a band of Mingo and/or Shawnee attacked the farm of James Graham, about 28 km (18 mi) below Arbuckle's Fort in present Summers County, West Virginia. It is likely that this was the same party that later attacked the fort.

The James Graham farmstead was first settled in 1770 when a small log cabin was constructed. A few years later a much larger, two-story log house was built, and the smaller cabin became the kitchen (Graham 1905:15). In his September 12 letter, Captain Stuart stated, "I was also alarmed at the same time with an account from James Graham . . . who was yesterday attacked by a party of Indians at this house who killed Walter Caldwell as he was shuting the door to keep them out, several other persons were killed and taken at the same time. I am told after the people got relief from Cap't. Jno. VanBibbers who lives in sight of the Grahams" (Thwaites and Kellogg 1912:80). This attack was also described by Captains John VanBibber and James Henderson; the first was stationed at VanBibber's Fort across the Greenbrier River from the Graham farm. According to VanBibber, "About Break of day this morning they attacked the house of James Graham which is situated within three hundred yards of the fort where they killed three and took one prisoner, and in about two hours afterwards a small Detachment of men which was going to the Assistance of some Adjacent Neighbours was again Attacked within two hundred yards of the Fort, where our men gave them Battle and sustained no damage only one man slightly wounded in the shoulder, what loss the Enemy sustained is to us unknown" (Thwaites and Kellogg 1912:78). According to Henderson, "This morning I Recd the following Acct from Andw Kinkead which is as follows—that he and Walter Caldwell was at James Graham's in GreenBrier River last wensday Night and a Thursday morning a litel before Day the Indians broke open the Dore upon them, and Shott Walter Caldwell & kild one of Jas. Grahams Children and Negro Fellos; and one of his Children is taken prisenor" (Thwaites and Kellogg 1912:79–80). Neither of these descriptions is very detailed, but they do suggest that the Indians observed that the Graham house door was open or at least vulnerable and attacked that point and killed and captured several people. These reports also describe a rescue party from the fort that engaged the Indian raiding party two hours later and took one casualty.

Fortunately, additional detail on the Graham house attack was provided by James Graham's grandson David in two publications (Graham 1899, 1905). David Graham (1899:90) stated that a few days before the attack Indians were sighted in the neighborhood, and an alarm had been sounded that "was heralded from home to home . . . [and] the whole community collected themselves to the fort [VanBibber's Fort]. . . . After being in the fort for a short time, possibly two or three days, and no further signs of the Indians . . . James Graham . . . proposed to those in the fort that, if some of the [militia] men would go and stay with him a night, he and his family would go over home. Accordingly, they did so." This proved to be a fateful decision for the Graham family, as the Indian raiding party had remained in the neighborhood undetected and kept watch on VanBibber's Fort from hidden positions.

According to David Graham (1899:90–91), "The assault was made in the after part of the night before daybreak. . . . The Indians made the assault by trying to force the door open, which they partly succeeded in doing. Thus aroused, Graham and his men placed the heavy bench and a tub of water against the door. . . . A man named McDonald (or Caldwell), who was assisting in placing the tub against the door . . . was shot and killed, the ball passing through the door."

Following this initial assault, Graham (1899:91) adds, "Thwarted in their effort in affecting an entrance into the house, the Indians next turned their villainous assault upon an outhouse or kitchen standing near the main dwelling. In this outbuilding slept a young negro man and two of the Graham children." The Indians successfully broke into this building, killed the African American man, whose name was Sharp, wounded the boy, John Graham, in the loft and later killed him "out in the yard" as "he would be a burdensome prisoner," and made captive the girl, Elizabeth Graham (Graham 1899:92). "While this bloody scene was going on in the kitchen, Colonel Graham [who was a militia lieutenant at the time] had gone upstairs [in the house] and was shooting through a porthole at the Indians in the yard as best he could. The men in the lower part of the house loaded the guns and handed them up to him and he did the shooting. About the time [the Indians] were trying to make the wounded boy stand up [in the yard] several [of the militiamen] huddled together and fired at the bulk [of the Indians]; when they suddenly dispersed. It is believed that one or more of the Indians were killed or wounded" (Graham 1899:92).

The Graham house assault illustrates several basic Native American tactics of irregular warfare. These include stealth, when they hid unobserved near VanBibber's Fort for two to three days; surprise, when they attacked the vulnerable Graham house door before daybreak; and finally, a desire to minimize casualties, when they broke off the initial attack on the house and switched to the undefended kitchen and then broke off the attack altogether when they came under heavy fire from the house. The decision of the raiders to observe, rather than attack, VanBibber's Fort also fits into their tactic of not attacking unless success and minimal loss of life were highly probable.

The raid illustrates the frontier settlers' most common defensive tactic of fortifying within strong buildings or forts when under threat. It also illustrates the cost of being careless: the house door being left vulnerable; Sharp and the children sleeping in the less defended kitchen; and especially the settlers underestimating their enemy in leaving VanBibber's Fort prematurely. In contrast, the raid would have been viewed as successful by the Indians. They managed to spread terror among the settlers, inflicted numerous casualties, took a war captive, and suffered no, or very few, casualties. James Graham was able to ransom his daughter back from the Shawnee in 1785 (Graham 1899:96).

Archaeology of the Graham Farm and Arbuckle's Fort

Archaeological investigations at the Graham homestead and Arbuckle's Fort, aided by the battle descriptions, were able to add new information on the battles and defensive architecture. Descriptions of the Graham house attack noted the presence of a detached kitchen at the farm but also stated that the kitchen was the original 1770 Graham residence. David Graham's (1899, 1905) description of the farm indicated that this kitchen was near the main residence occupied at the time of the raid. Given the importance of both structures to the events of the raid, a program of archaeological survey and testing was designed to locate the kitchen and gain a general understanding of the use of the Graham house yard (Figure 1.1).

Archaeological survey of the Graham house proper, which is still standing, consisted of shovel test pit excavation and a very limited program of metal detecting. These tests produced strong evidence, particularly in the distribution of creamware and pearlware ceramics, that renovations of the house had reversed its front to rear orientation during the early to mid-nineteenth century, with the original front door facing north toward Graham's Ferry. The distribution of period kitchen/dining artifacts in the shovel test pits in the original yard area suggests that the detached log kitchen stood about 9.1 m (30 ft) to the southeast of the main residence. The discovery of a deeper, buried cultural strata and possible foundation stones in this area led to an expanded excavation that located a stone foundation and cellar of an eighteenth-century building. The substantial nature of these features and the presence of eighteenth-century ceramics, glassware, and wrought nails suggest that this building was likely the Graham kitchen, the site of the bloodiest portion of the attack. While no definite battle-attributed artifacts were recovered, the proximity of this building to the south door of the main house increases the probability that this was the focus of the initial predawn Indian attack.

Arbuckle's Fort is the most extensively excavated, and best understood,

Figure 1.1. Archaeology at the James Graham house. (Courtesy of W. Stephen McBride and Kim A. McBride)

of any of the Greenbrier Valley fort sites. Here archaeology has found and exposed the footprint of the primary bastioned fort stockade wall, 33.5 × 33.5 m (110 × 110 ft), two shorter stockade lines (one with a small bastion or redan), the stone foundation and chimney of a blockhouse, a blacksmithing area, four cellar pits, and several refuse pits and associated post molds (Figures 1.2 and 1.3). One question posed by these excavations concerns the nature and function of the two shorter stockade lines: are they part of an earlier smaller fort, or could they be secondary lines of defense within the

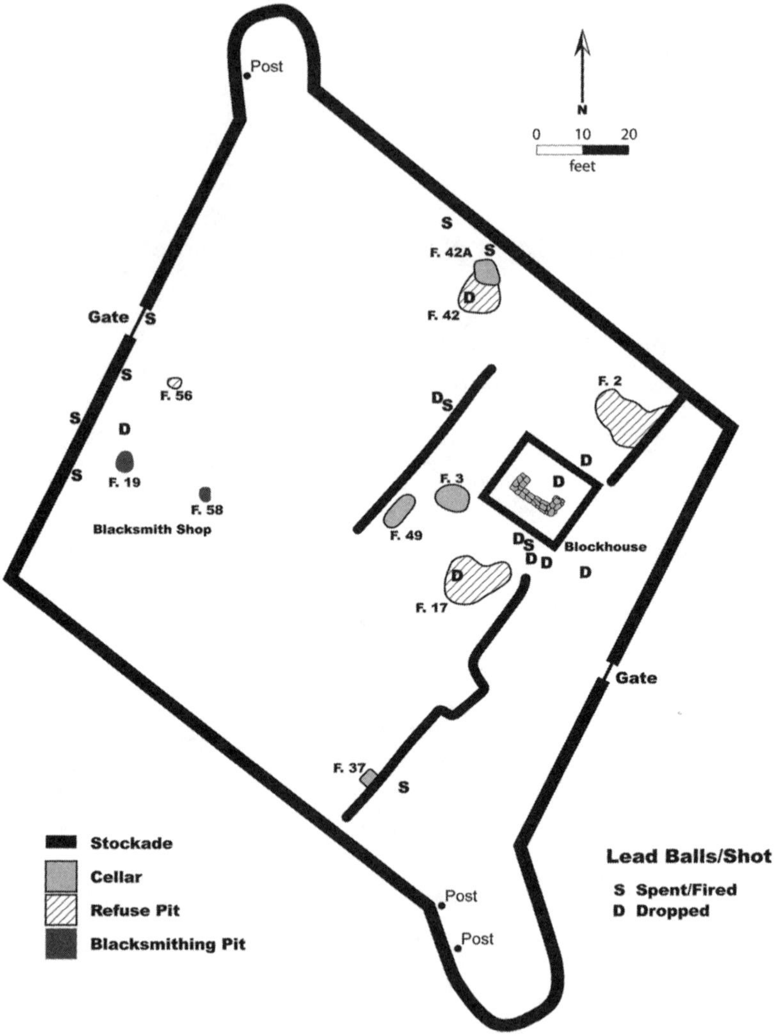

Figure 1.2. Distribution of lead balls and shot at Arbuckle's Fort. F. = feature. (Courtesy of W. Stephen McBride and Kim A. McBride)

larger fort? The distribution of arms-related artifacts found in test units and by metal detecting helps answer this question and gives insights into the 1774 and 1777 Indian attacks.

The distribution of spent or fired round-ball ammunition, interpreted as coming from Indian attackers, indicates firing on both the northwestern and northeastern main walls of the fort, but also firing on the two shorter stockade walls and the blockhouse (see Figure 1.2). Although it is possible for the blockhouse to have been fired on over the main stockade, it would

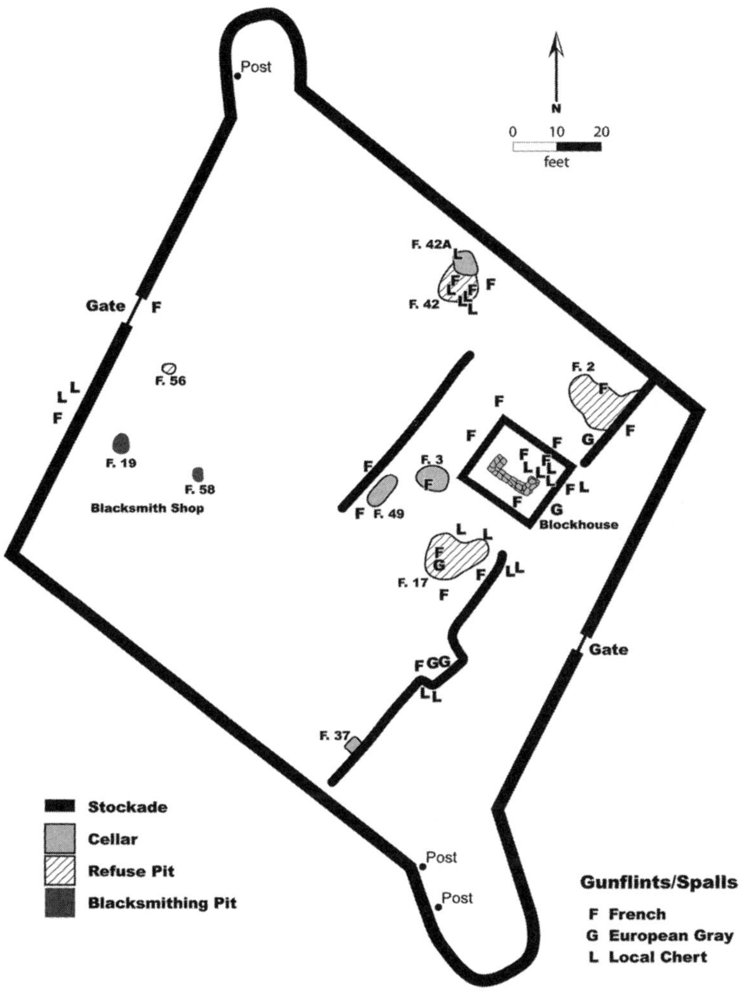

Figure 1.3. Distribution of gunflints at Arbuckle's Fort. F. = feature. (Courtesy of W. Stephen McBride and Kim A. McBride)

have been nearly impossible to hit the short walls if the main stockade was present unless it had been breached. That seems very unlikely given the descriptions of this attack. A better explanation for this pattern is that the short walls were part of an earlier fort and that these fired balls probably represent the 1774 attack, which occurred from the southeast and northwest, whereas the balls at the outer stockade walls represent the 1777 attack, which came mostly from the north.

Further evidence for this conclusion is found in the placement of dropped (unfired) round balls, shot, and especially the gunflints that often serve to identify a firing line, or at least a location from which firing of a musket took place. In this instance, the dropped balls (.40–.50 in) and shot (.28–.32 in) suggest militia firing positions at the blockhouse, along the northwestern short stockade, and along the main northwestern and northeastern stockade walls. The lack of dropped balls in the two bastions is surprising but may be explained by limited excavation in these areas.

Three types of gunflints have been recovered at Arbuckle's Fort: honey-colored French gunflints, medium-gray European gunspalls, and gunspalls made of local, gray Hillsdale chert. The French flints and those of local chert are concentrated in both of the short stockades, the blockhouse, the northeastern structure (as defined by a cellar pit), and the northwestern main stockade. The European gray gunspalls, in contrast, are concentrated only along the southeastern short stockade, particularly in the redan (see Figure 1.3). The presence of gunflints, particularly the European gray gunspalls, and dropped ammunition along the short stockade walls suggests that they were heavily defended, making it unlikely that they were mere secondary lines of defense.

The distribution of fired and dropped ammunition and gunflints/gunspalls at Arbuckle's Fort suggests that the 1774 Indian raiding party advanced up Muddy Creek, to the west, and Mill Creek, to the east/southeast, in an effort to envelope the fort. They then fired on the fort's northwestern and southeastern walls, and likely the blockhouse. The defending militiamen returned fire from the stockade walls and the blockhouse. With trained militiamen defending the fort, the Indians likely broke off the attack because victory seemed uncertain.

The 1777 attack could have come from either or both Muddy and Mill Creeks, but because the main firing came on the northwestern and northeastern walls of the fort and the blockhouse, it is more likely that the Indians approached from just the Muddy Creek side and encircled the northern half of the fort. Again, like 1774, the Indians probably broke off the attack when success seemed uncertain.

National Warfare: Donnally's Fort

The 1778 attack on Donnally's Fort was on a much larger scale than any of the previously mentioned battles and involved the largest number of combatants of any colonial or Revolutionary War action in the Greenbrier Valley. It possibly involved upwards of two or even three hundred Indians (Wyandot, Mingo, and likely Shawnee) under the leadership of the Wyandot Half King in a contest with about 98 militiamen and officers commanded by Lieutenant Colonel Andrew Donnally and Colonel Samuel Lewis (Kellogg 1916:70–73; Stuart 1971 [1833], 1820). The battle occurred at Donnally's

Fort, which was built as a residence about 1771 in Rader's Valley, Greenbrier County, and then fortified between 1774 and 1776 (Dayton 1942; McBride and McBride 2006). Andrew Donnally was a prominent settler of the Greenbrier Valley and held numerous political and militia positions including captain (1776–1777), lieutenant colonel (1778+), and county lieutenant (1781) (McBride and McBride 2006:22; Summers 1929:222–232). Because of the size and significance of this battle, Donnally's Fort is the best described of any Greenbrier fort and reportedly consisted of a bastioned stockade that surrounded or intersected Andrew Donnally's log house and kitchen (Bickley 1993; Royall 1991 [1826]; Stuart 1971 [1833], 1820).

The battle at Donnally's Fort was the culmination of a larger Indian campaign that began at Upper Sandusky, Ohio. The war band crossed the Ohio River near the mouth of the Kanawha River and laid siege to Fort Randolph, at present-day Point Pleasant, West Virginia, before moving up the Kanawha River and finally crossing into the Greenbrier Valley (Kellogg 1916:64–65, 70–73, 82–83; Stuart 1971 [1833], 1820). The goals of this campaign likely included punishing the garrison at Fort Randolph for their earlier murder of Shawnee chief Cornstalk and three companions, and spreading retaliatory death and destruction in the Greenbrier Valley (Downes 1968:208–209). Participating Virginia soldiers and militiamen estimated that the Indian force consisted of 200 to 300 warriors (see Arbuckle 1778; McKee 1778; Stuart 1971 [1833]). The Moravian missionary Rev. David Zeisberger (1778), writing from the Christian Delaware town of Coshocton, reported that "the Half King [Dowyentet] of the Wyandots with all his men and Mingoes with him, a hundred in all, are gone to the fort at Kanawha [Fort Randolph]." Zeisberger also believed that only half this party continued the campaign after the attack on Fort Randolph. Other scholars believe that the Shawnee were also involved in this campaign and may have joined the Wyandot and Mingo after they left the Delaware town (Calloway 2007:68; Downes 1968:208–209; Wood ca. 2014). We do know that after an unsuccessful siege of Fort Randolph, a large multitribal Indian raiding party moved toward the Greenbrier settlements.

Fearing for the safely of the settlers, Captain William McKee, commander of Fort Randolph, sent two volunteers, Philip Hammond and John Pryor, to pursue and pass the Indian party and warn the Greenbrier settlers. Fortunately for the settlers, Hammond and Pryor arrived at Donnally's Fort ahead of the Indians on the evening of May 27, 1778 (Arbuckle 1778; McKee 1778; Stuart 1971 [1833], 1820). The alarm was quickly spread throughout the region, and settlers quickly fled to their neighborhood forts, including Donnally's, which was garrisoned by Lieutenant Colonel Donnally, one or two militia officers, and 27 militiamen. Sixty women and children and at least one slave (Dick Pointer), but probably more enslaved African Americans, were also present (Kellogg 1916:72; Stuart 1820).

On the evening before the battle, the Indians encamped on Brushy Ridge to the northwest of the fort. The next morning (May 29) they left the ridge and moved into the Rader's Valley north of the fort and then advanced along the run to the fort (Hardesty 1883:355, 395). As the Indians moved toward the fort, they killed a man gathering firewood near the creek. Then, according to battle participant Captain Stuart (1820), "The Indians commenced their attack about day light when all in the fort were fast asleep except the centry, Phil Hammon and an old negro [Dick Pointer]. The house composed one part of the front of the fort and was double. Hammon and the negro was in the kitchen and where a hogshead of water [was] against the door. The Indians had laid down 3 guns at a stable about 50 yards from the house and made their attack at the kitchen door with their tomhawks and war clubs when Hammon let the door open and killed the Indian on the threshold . . . the negro had a musket loaded with swanshot and . . . fire[d] away among them. . . . The firing of Hammond awakened the people in the other end of the house and upstairs . . . who fired out of the windows so briskly upon the enemy that . . . 17 lay dead in the yard."

After this unsuccessful frontal attack on the house/kitchen part of the fort, the Indians made an attempt to get under Donnally's house and "wounded a man through the cracks in the floor" but were beaten off by "the whites slipping loose planks to one side, and shooting down among them" (Cartmill ca. 1840; Taylor 1829). Following this, the Indians surrounded the fort and placed it under siege. Two militiamen were killed while attempting to reach the fort, and another was "shot in a bastion in the fort" before a rescue party of 68 militiamen and officers from Fort Savannah (near present-day Lewisburg) under Colonel Lewis and Captain Stuart successfully averted an ambush on the main road and entered the fort at 2:00 p.m. (Stuart 1971 [1833], 1820). The siege continued into the evening when an Indian "drew near the fort and called out in English that he wanted to make peace. We invited him to consult on the terms, but he declined our civility" (Stuart 1971 [1833]:32). Later that evening the Indians departed after retrieving eight of their 17 dead. Some Indians returned to Ohio and others dispersed into smaller parties and continued raiding (McBride and McBride 2014:14).

The Donnally battle illustrates the Indians' use of early morning surprise and a quick assault on what they perceived to be a possible weak point—the kitchen door. It differed from the 1774 and 1777 battles at Arbuckle's Fort, however, in that after the initial assault failed, the Indians settled into a siege. This difference is certainly related to the much larger number of Indians in this battle. Nevertheless, their failure to ambush the rescue party of militia from Fort Savannah probably sealed the ultimate outcome of the battle, although the Indians did make a half-hearted attempt to negotiate a positive outcome. The loss of at least 17 Indian men versus four on the militia/settler side certainly makes this attack a failure for the Indians. The militia and settlers utilized their usual strategy of forting up, the success of which was greatly aided by the warning of Captain McKee's scouts. The plan and action of the rescue party to leave the road, thus avoiding the Indian ambush, and reinforce the fort's defenders, were also critical moments in this battle.

Archaeology of Donnally's Fort

Descriptions of the battle and fort, which are by far the best in the region, were critical to developing a program of archaeology that had as its major goal to discover the design and position of the fort's stockade and bastions and the location of the Donnally house/kitchen within that complex. Obviously, the discovery of the fort's footprint would also help to better understand the battle.

The most important descriptions of the fort and the battle were those of the two battle participants John Stuart (1971 [1833], 1820) and Dick Pointer (Royall 1991 [1826]). According to Stuart (1820), "the house formed one part of the front of the fort and was double [with] the kitchen making one end of the house." Pointer, in turn, reported that "Donnally's house made part of the fort, the front of it forming a line with the same, the door of the house being the door of the fort" (Royall 1991 [1826]).

Stuart (1971 [1833], 1820) also noted that one man was killed in a bastion of the fort. These descriptions state that the house was likely intersected by the stockade at the front of the fort. Accordingly, if the location of the house could be determined, the intersecting stockade should be located. This is exactly what happened. Once a double-stone chimney and a few stone piers attributed to the house were identified, trenching was carried out to locate the stockade. This work resulted in the identification of a two-bastioned stockade of 27.7 × 25.3 m (91 × 83 ft), a 20.4 × 12.2 m (67 × 40 ft) double pen house on stone piers with a 3.5 m (11.6 ft) central double (two-hearth) chimney, a cellar to the rear of the Donnally house, and two pits/privies in the western corner of the fort. The two bastions were located on the north and south corners of the fort. Each was 3.05 m (10 ft) square. Scaffold or floor support posts were uncovered in the southwestern bastion. A 5.03 m (16.5 ft) gap, which is certainly the fort gate, was discovered along the northeastern wall (Figure 1.4).

The archaeological evidence from Donnally's and Arbuckle's Forts suggests that the militia/settlers' concept of a larger neighborhood fort was a two-bastioned wooden stockade roughly 27.4–30.5 m (90–100 ft) square.

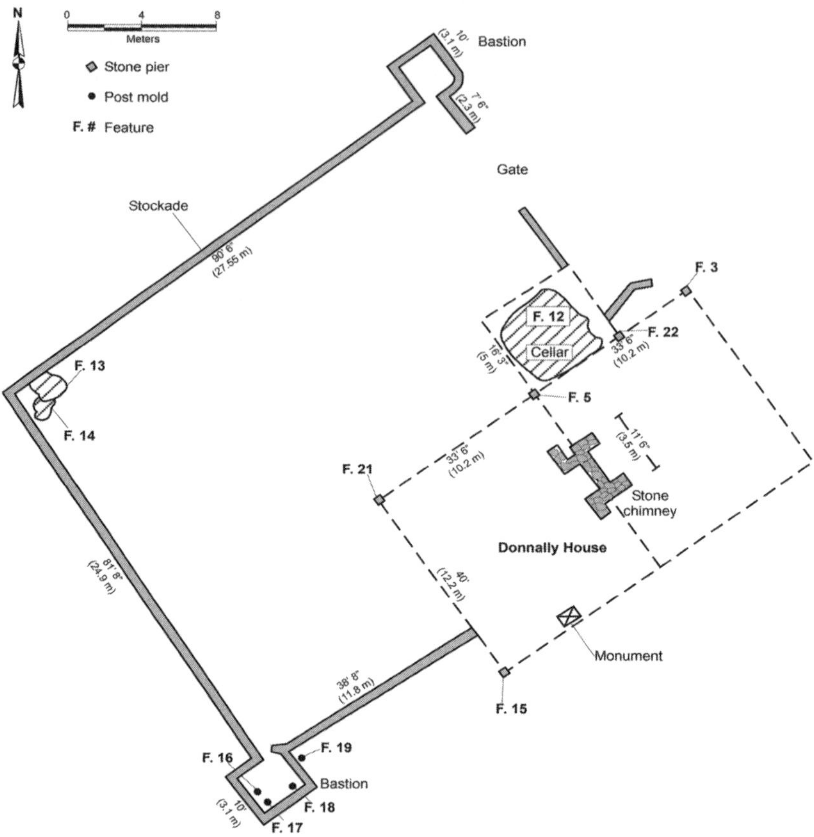

Figure 1.4. Layout of Donnally's Fort. (Courtesy of W. Stephen McBride and Kim A. McBride)

This plan was viewed as both large enough to hold a garrison and neighborhood refugees and strong enough to withstand an Indian attack.

The archaeological evidence also allows us to reconstruct the action taking place in parts of the engagement area. Because the northern hearth of the double chimney identified at Fort Donnally was the wider (2.3 m [7.5 ft] versus 1.7 m [5.5 ft]) of the two hearths, it was certainly the kitchen fireplace. With reference to battle history, the northern (10.2 m [33.5 ft]) pen or half of the house in which the kitchen was located was the locus of the main Indian assault. Although no definite battle artifacts were found during the excavation, two dropped 0.45–0.50-inch lead balls were recovered near the Donnally house that give evidence to the militiamen's weaponry. Following this frontal assault, the Indians enveloped the fort, shot one militiaman in a bastion, and initiated a siege. The position of the fort and its gate relative to the creek and road suggests that the militia rescue party from Fort Savannah mostly likely approached the fort from the northwest rather than the southwest. The latter approach would place them closer to, or on, the road, farther from the gate, and in the low-lying creek bottom in a position making them vulnerable to attack.

Conclusions

The historical and archaeological investigations of the three Greenbrier Valley battle sites have provided information on Indian and settler modes of warfare, on the physical layout of three sites of conflict between Native Americans and settlers/militia, and on the battlefield landscape over which the actions took place. The attacks on the Graham farmstead and Donnally's Fort illustrate two Native American tactics: use of early morning surprise assaults on suspected weak points—doors in both cases; and rapid shifts in tactics—surround and siege at Donnally's, and attacking the weaker detached kitchen at Graham's farm—when an initial surprise attack failed. These shifts suggest previously agreed-to tactical flexibility on the part of the raiders and the existence of good leadership, both of which worked to prevent high casualties as would have likely resulted from continued frontal assaults. The siege of Donnally's Fort also reflects the possibilities open to a war band when larger numbers of warriors are involved.

The successful defense of the doors or points of weakness by the militia/settlers/enslaved man at both sites was certainly aided by earlier warnings of Indians by scouts/spies. As a result, two men were on guard at Donnally's Fort, and the Grahams were able to feel secure, at least initially, by moving the family to VanBibber's Fort and later brought extra militiamen to their house. Advance warning also led to militia reinforcements being sent from Fort Savannah to help break the siege at Donnally's Fort.

Little is known historically about the two attacks on Arbuckle's Fort, but the general lack of local panic on the part of the settlers and militia suggests that these battles did not involve large raiding parties on the scale of Donnally's. Documents do suggest that militia and settlers were warned before the 1774 attack at Arbuckle's, and a similar warning, particularly given the earlier attack on the Graham house, may have contributed to the failure of the 1777 raid. Such alerts likely reduced the Indian warriors' chance of surprise.

Archaeological evidence from Arbuckle's Fort indicates that the Indians possibly fired on two to three sides of the fort in 1774 and two sides in 1777. This would suggest an attempt to partially surround the fort—perhaps in a half-moon formation—during these battles. Archaeological data also strongly suggests that Arbuckle's Fort was enlarged and improved between the two battles. The improved, later, two-bastioned fort makes the second attack somewhat perplexing as smaller raiding parties rarely attacked strong fortifications. Perhaps the war party was somewhat larger than suspected, or perhaps they were just probing the fort for weaknesses and then were moving on.

The more detailed battle descriptions of Donnally's Fort and Graham's farm greatly aided archaeology in discovering the architectural plan of these two sites. Archaeology also expanded our understanding of the evolution and spatial aspects of the battles. At Donnally's, evidence of a two-bastioned stockade and the Donnally house was discovered and exposed. At Graham's Farm the position of the external kitchen in the farm yard behind the main house was identified. At Donnally's Fort the position of the Donnally house within the line of the stockade served to diminish the strength of the stockade, the two front doors creating points of weakness, the doors probably being no stronger than those of a settler's home. In contrast the full stockade that enclosed Arbuckle's provided a stronger defensive line.

These sites, even the Graham house, illustrate the success of the militia/settler strategies and tactics of spying and forting up, and in the case of Donnally's, the utility of the Ohio River forts as an early warning system. Although Graham's kitchen fell with three casualties, the main house withstood the attack, so it was only a partial victory for the Indians. Whereas the Arbuckle's Fort battles were victories for the militia/settlers, they were not significant defeats for the Indians, as they had few if any casualties and moved on, without pursuit, and likely attacked more vulnerable locales. The Donnally's Fort battle was a more significant defeat both tactically and strategically because the Indians suffered moderate losses, at least 17 killed, and ended a large-scale campaign that had lofty goals with a defeat.

Acknowledgments

The Summers County Historic Landmarks Commission, under the guidance of Stephen Trail and administrative capabilities of Myra Ziegler, has supported the research from its inception in the early 1990s, including administration of grants from the West Virginia Division of Culture and History; the West Virginia Humanities Council; and the West Virginia Budget Digest, Fairs and Festivals, and Historic Preservation funds. This research was only possible with the support of the Greenbrier Historical Society and local landowners: in the case of Arbuckle's Fort, the Dan Clay and Diana Shay family; for Donnally's Fort, the John and Drema Johnston family; and in the case of the Graham house, the Graham House Preservation Society.

2
"Foot Jägers Forward!"
Johann Ewald, Petite Guerre, *and the Archaeology of the Battle of Cooch's Bridge*

Wade P. Catts

In the late summer of 1777, the people of the upper Delmarva Peninsula and adjoining southeastern Pennsylvania witnessed the start of the military campaign that resulted in the capture of Philadelphia on September 26, 1777, by the British army. Two years of warfare had preceded the Philadelphia campaign, with much of the principal military action on land occurring in New England, New York, and New Jersey. British setbacks occurred in December 1776 (First Battle of Trenton) and in early January 1777 (Battle of Princeton). These reversals resulted in British forces wintering in New York City and near New Brunswick and Perth Amboy, New Jersey, and American forces taking up winter quarters around Morristown, New Jersey.

During May and June 1777, the inability of General Sir William Howe's British army to force a decisive battle or outmaneuver General George Washington's Continental army in New Jersey led to the movement by sea of the balance of the British army from Staten Island to Elk Neck in Maryland. Numbering approximately 15,000 fighting men, the British army was transported by the Royal Navy up the Chesapeake Bay and landed at the Head of Elk on August 25, 1777. Its intent was to advance overland to capture Philadelphia, the capital of the fledgling United States.

From August to December, American and British forces battled their way across the landscapes and waters of northeastern Maryland and the lower Delaware Valley. In a series of engagements—Cooch's Bridge (September 3), Brandywine (September 11), the Battle of the Clouds (September 16), and Paoli (September 20–21)—the British maneuvered and fought their way to the colonial capital of Philadelphia, a prize that was captured on September 26 (McGuire 2006).

The capture of the city by the British, however, was not strategically decisive. Washington's army assaulted a detachment of the British army at Germantown (October 4) and nearly succeeded in inflicting a significant defeat on them. Holding Philadelphia proved to be somewhat problematic. American forces had sunk obstructions and strongly fortified the Delaware

River preventing access to the city by the Royal Navy, and the small Continental and Pennsylvania state navies augmented these defenses. Chief among the fortified posts were Fort Mercer (or Red Bank) on the east side of the Delaware River and its sister fortification, Fort Mifflin, on the Pennsylvania side of the river. An unsuccessful Hessian assault on Fort Mercer (October 22) was followed by the siege of Fort Mifflin (November 10–20). Only after Mifflin was pulverized into submission was the river opened for naval traffic, and most importantly, badly needed supplies for the British forces in Philadelphia entered the city (McGuire 2006, 2007).

Howe made one final attempt to engage Washington's army in battle at White Marsh in early December. Following that brief fight, both armies retired to winter quarters, the British into Philadelphia, and Washington's Continentals to Valley Forge (December 19). At Valley Forge, the Continental army created a hut city and, for the first time in the war, began standardized training. The winter camps were occupied by both armies until June 1778, when the British abandoned Philadelphia and retreated overland to New York, fighting a principal engagement at Monmouth in late June.

The Battle of Cooch's Bridge, Delaware (sometimes known as the Battle of Iron Hill by American participants), fought on September 3, 1777, was the first significant engagement of the Philadelphia campaign (Catts et al. 2014; Cooch 1940). Described by participants as severe, sharp, and bloody, the battle represents a small unit action of the American War for Independence. The "British" unit most heavily engaged was the Hessian Field Jäger Corps, one of whose captains, Johann Ewald, excelled at small unit fighting and was later acknowledged as one of the best practitioners of *petite guerre* (irregular or asymmetric warfare) of the national wars of the eighteenth century. The British unit's opponents at Cooch's Bridge were an ad hoc formation of American "light infantry," composed of Continentals, militia, and volunteers under the command of Brigadier General William Maxwell. This chapter examines the battle of Cooch's Bridge from the archaeological data, serving as an example of contrasting military formations meeting in an asymmetric small unit action within a larger conventional setting.

The Battle of Cooch's Bridge

At 5:00 a.m. on the morning of September 3, 1777, the British army vanguard advanced from the Head of Elk toward Aiken's Tavern (Figure 2.1). This advance guard comprised troops of the Hessian Field Jäger Corps, two battalions of British light infantry, and two British grenadier battalions, followed by most of the remainder of the army. Contemporary British and German accounts note that the Hessian advance guard reached Aiken's Tavern

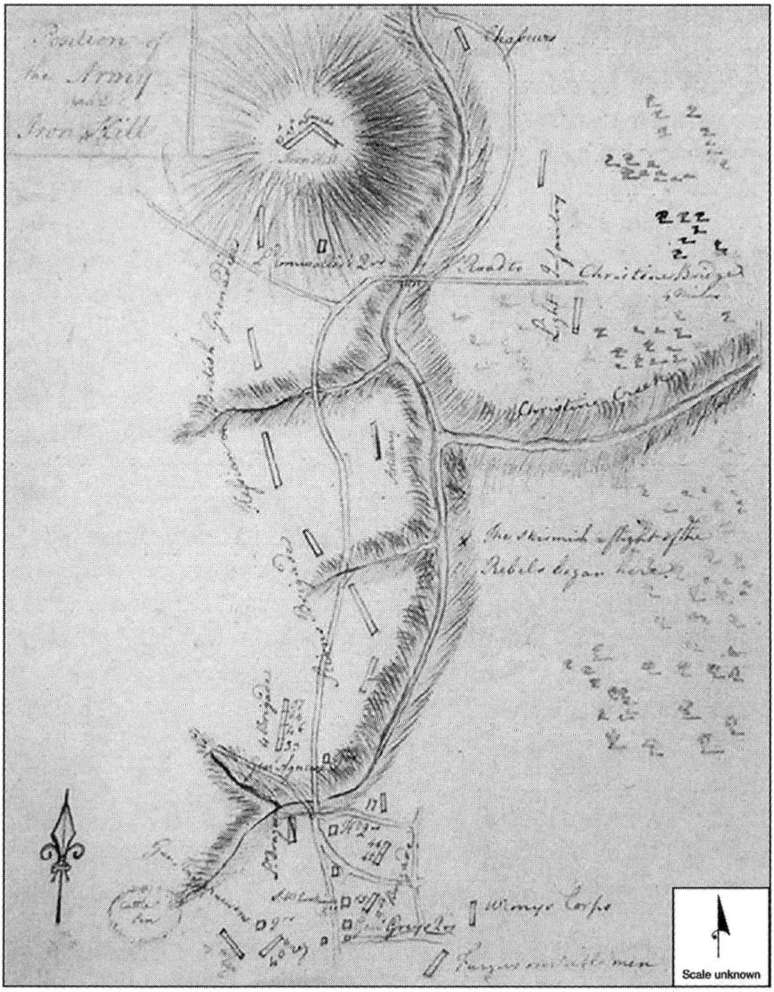

Figure 2.1. Unpublished manuscript map by Major John André, "The Position of the Army at Iron Hill." (Courtesy of the William L. Clements Library, University of Michigan)

where it was determined that the American troops had withdrawn along the road to Newark (present-day Route 896). The Hessian advance guard was under the command of Captain Ewald. Ewald advanced from Aiken's Tavern northward toward Cooch's Bridge with six mounted jäger. Around 8:00 a.m. at about 0.8 km (0.5 mi) north along the road, Ewald's command was ambushed by point-blank musket or rifle fire from a hedge, and all six of his men were either killed or wounded (Ewald 1979:77).

Ewald attacked this American force with his company but encountered a second American light infantry detachment in a woods and became heavily engaged. Maxwell's orders were to engage the enemy, force it to deploy, fall back to another defensive position, and delay the enemy again. The American light infantry, established only one week before the battle, had no intention of fighting a pitched battle against the advancing British army, but sought only to delay.

Lieutenant Colonel Ludwig von Wurmb, commander of the Hessian Field Jäger Corps, reinforced Ewald. He deployed his three companies in line, perpendicular to the Newark road (Retzer and Londahl-Smidt 1998). No sooner were the Hessians deployed than the American light infantry attacked the center of the line. The battle here between the jäger and the light infantry was spirited, but the Americans eventually quit this position, or they were driven from it, according to Hessian accounts. General Howe's German aide, Friedrich von Muenchhausen, observed American light infantry at this time as "behind trees, firing at our advancing jägers, then retreating 20 yds [18 m] behind the next tree, then firing again" (Muenchhausen 1974:28). Royal Artillery officer Francis Downman commented that the "rebels fired as usual from the fences and trees" (Whinyates 1898:156). After falling back for the first time, the Americans formed a line in the woods and faced the Hessians again. It is possible that Maxwell was fighting from a series of prepared or established lines, or that he was reinforcing his detachments with other light infantry units. Eight days later, Maxwell's delaying action at Brandywine was fought in a similar fashion (Harris 2014; McGuire 2006; Smith 1976).

Lieutenant Colonel Wurmb reported that the Americans in the second line "defended themselves obstinately" but were outflanked in hand-to-hand fighting with the Hessian right flank company under Wrede (Burgoyne 1987, 1995, 1997; Retzer and Londahl-Smidt 1998). It appears from the accounts that the American troops not only defended themselves but may have seriously pressured the jäger and nearly forced them to give way. Wrede's company fought the Americans with their short swords, or hangers. Later, Ewald wrote in his *Treatise on Partisan Warfare* (published in 1785) that if one is forced to retreat and is unable to do so because of the closeness of the enemy, he should not wait for the enemy to attack but instead attack the enemy with his bayonet. In this case, Ewald wrote, "the jäger on foot can draw their hunting knives . . . this is what Captain von Wrede had his jäger on foot do in the affair at Cooch's Mill in Delaware" (Selig and Skaggs 1991:124).

By this time, the sound and intensity of the musketry were steadily increasing, so that Howe determined to send in both battalions of British light infantry in an attempt to outflank Maxwell's line. The 1st Light Infantry Battalion moved to the right (or east) to cut off the American line of retreat, but as one British officer complained, the movement of the battalion was "prevented by an impassable morass, which the guide was not acquainted with" (Stirke 1961:168). It is likely that the "impassable morass" is in the present location of Sunset Lake, a body of water created as a mill pond in the nineteenth century and absent at the time of the battle.

The 2nd Light Infantry Battalion moved to the left or west. It too encountered a swampy area but continued its advance. The battalion apparently went too far to the left and engaged the New Castle County militia under Colonel Thomas Duff. According to Muenchhausen, the militia fired only one round then left the field in haste and disorder. The fire of the militia may have been sufficient to slow the 2nd Light Infantry Battalion down, keep them out of this phase of the battle, and cost Colonel Duff his horse (Pennsylvania Gazette, 10 September 1777). Duff's battalion was probably posted along the southern margin of Iron Hill, on the main Philadelphia road.

During the time that the second American line held, the firing was reported as intense, and casualties on both sides were mounting. Like the first line, this one was established in a woods. The action here seems to have been heated. Lieutenant Colonel Wurmb was described as "continuously in front of the jäger, encouraging them in every way, both by actions and by words." Outflanked to their left by Wrede's attack and engaged in hand-to-hand combat, the American light infantry in the second line finally shot themselves out of ammunition and were forced to withdraw again (Muenchhausen 1974:28).

The American light infantry withdrew and formed a third defensive line, this time to the east, across Christiana Creek at Cooch's Bridge. Farther south, Howe, concerned that the firing was still heavy and that his light infantry battalions were not yet engaged, sent forward both battalions of British grenadiers and two cannon to support the jäger. In the combined force of Hessian jäger, British light infantry and British grenadiers, Howe was prepared to commit approximately 3,000 men to the fight, a force several times larger than Maxwell's entire corps.

The third American battle line at Cooch's Bridge held for a time, but the jäger, now reinforced by the 2nd British Light Infantry, forced the Continentals back from the bridge. The jäger's small one-pounder and battalion's three-pounder guns apparently came into play during the fight at the bridge. These guns were loaded with grapeshot, which was used to deadly effect. Artillerist Downman noted that he saw "a corporal and five men lying near together, killed by grape shot" (Whinyates 1898:156). The British light infantry stormed across the bridge, taking a few casualties in the process.

The engagement had lasted about two hours. Maxwell's troops, their ammunition exhausted, facing a large enemy force with artillery and unsupported by the rest of the American army, withdrew. The troops moved to the east and into thick woods, then toward the village of Christiana Bridge, about six miles away. The British controlled the field, holding the bridge and the main road to Philadelphia.

Casualty reports for the engagement are difficult to interpret. British and German accounts indicate that their losses included one or two Hessian jäger killed, 15 Hessian and four Anspach jäger wounded, one British noncommissioned officer killed, and three officers and 19 enlisted men wounded. British losses were mostly confined to the 2nd Light Infantry.

American casualties as reported by the British ranged from 20 to 50 or more. American wounded were carried from the field, but the dead remained to be buried by British pioneers. Lieutenant Gilbert Purdy of the British Corps of Guides and Pioneers noted that his unit buried 24 dead Americans on the field, including many officers (Purdy 1777–1778). Based on information derived from muster rolls, pension records, and other contemporary sources, the heaviest American losses apparently fell in the light detachments drawn from Conway's Brigade, Weedon's Virginia Brigade, and Hartley's (Wayne's) and Humpton's Pennsylvania Brigades.

Cooch's Bridge was the first time that the Hessian Field Jäger Corps's several companies fought together as a single military formation. The battle was also the first time that the various units of Maxwell's light infantry corps had fought together. The tactics employed by the opponents reflected their mission, training, and degree of cohesiveness as military formations. The Cooch's Bridge battlefield has been archaeologically investigated, and the recovered artifacts contribute significantly to new interpretations and understandings of the battle, particularly an understanding of scale. The engagement epitomizes many of the short but bloody fights that occurred during the American War for Independence. It represents a class of engagement much smaller, in both terms of the numbers engaged and extent of geographic area, than the larger and better reported principal battles of the war.

Military Formations and Tactics

At Cooch's Bridge two contrasting military formations fought against each other. These were the recently established American light infantry corps and the British army. Each of these formations was composed of smaller military units, and each was armed and organized somewhat differently. The differences in armaments among the two armies, as seen by archaeologically recovered lead shot, leaves a distinctive archaeological signature that allows their respective battlefield activities to be studied.

American Light Infantry Corps

On August 28, 1777, from his headquarters in Wilmington, Delaware, Washington ordered that "a Corps of light infantry is to be formed, to consist of one field officer, two captains, six subalterns, eight sergeants, and 100 rank and file from each brigade" (Washington 1777a). The corps was intended to replace Daniel Morgan's rifle battalion, which Washington had detached from the main army to reinforce the northern army under General Horatio Gates, which was facing the advance of Lieutenant General John Burgoyne's army in the Hudson Valley. The nine brigades forming the main army were ordered to furnish "100 men that are good marksmen to form a Light Infantry Corps . . . as this is meant only for a temporary establishment & as the utility will depend upon the goodness of the men and Officers for such a Service, the Genl. desires the Commanding Officers of Regiments to send none but such as may be depended upon" (Greene 1980 [1777]:147).

In contrast to the British army (see below), the American army was composed of long-term Continental regiments from the states, state-raised militia companies serving for short terms (one to three months generally, sometimes longer), and volunteers. Many of the state regiments had followed the British pattern and established light infantry companies, but in 1777 this practice was inconsistent among the states. Thus some state regiments, notably New Jersey and Virginia, had "light companies," and some of these companies were rifle armed, and others were not.

The role of the light infantry corps was to gather intelligence, harass the British in their movements, and act as an advanced guard for the Americans (Wright 1926). The creation of the light corps was an early use of regular troops in a large formation specifically designated as light infantry in the history of the American army. Under the command of Brigadier General Maxwell of New Jersey, the chosen men were intended to be marksmen, and the corps was a "combined arms" formation including soldiers armed with rifles as well as muskets. In a letter to General Gates, Captain Walter Stewart wrote that the light corps was composed of soldiers armed "mostly with rifles" (Stewart 1777). In addition to musket-armed Continentals drawn from Virginia, Pennsylvania, North Carolina, and New Jersey regiments, the American light infantry corps that fought at Cooch's Bridge included a large proportion of riflemen, many of whom were volunteers from the militia companies drawn from the counties of Lancaster, York, North Hampton, and Cumberland in western Pennsylvania. These Pennsylvania militia company volunteers were attached to the light corps because they were armed with rifles (Cummings 1832). Militia companies from New Castle County, Delaware, were also attached to the light corps (Catts et al.

2014). Maxwell's light corps was thus a sort of mixed-arms formation, consisting of musket men, riflemen, and some mounted dragoons, drawn from Continental and militia sources.

The light corps was formed only one week prior to the battle at Cooch's Bridge. It served again at Brandywine on September 11 and at the Battle of the Clouds on September 16 and was disbanded, and the men returned to their respective regiments at the end of September. The fight at Cooch's Bridge was thus the first action of a military formation that was composed of many soldiers who had no prior experience serving together or any sense of military cohesion.

Although the light corps was itself brand new, many of the officers and men had previous combat experience. Some of the Continental officers in the corps had seen service during the French and Indian War, including Maxwell and Major Francis Gurney of Philadelphia. Others had been involved in the War for Independence for at least one year, including Virginians Lieutenant John Marshall (later chief justice of the United States) and captains Charles Porterfield and William Heth, and New Jersey captain Archibald Dallas. Even the Delaware militia included men who had served in the Flying Camp of 1776 or during the Trenton and Princeton Campaign earlier in the year (Catts et al. 2014).

The lack of unit training and cohesion, so critical to a successful offensive military action, was the Achilles' heel of the corps and the principal factor dictating how Maxwell's corps performed in combat. The tactics chosen to be used at Cooch's Bridge, and at the Battle of Brandywine eight days later, reflect the recognition by the corps's officers that they were deficient in cohesion. Instead the corps relied on establishing set defensive lines or positions while engaging the advancing enemy. In several instances, the tactics called for ambushing the lead enemy formations. Jointly, these tactics were designed to force the British to deploy into battle lines to dislodge the light corps. The Americans would then retreat before being overrun to another already prepared battle line. In his pension application light infantryman Sergeant Samuel Patton of Hartley's Additional Regiment explained the tactic, testifying that "as we retreated a part of us hid in ambush until the enemy came in gun shot, then [we] fired, retreated and loaded continuing to retreat and load until we met the main army" (Patton 1830). While being forced from one position to another, the light corps was able to slow the movement of British forces, inflict casualties, and preserve its own force.

Hessian Field Jäger Corps

Opposing the American light infantry corps at Cooch's Bridge was a military unit known as the Hessian Field Jäger Corps. Howe established the Hessian Field Jäger Corps in late June 1777 in preparation for the summer campaign. Howe, an early practitioner of the use of light infantry tactics and formations, realized the usefulness and abilities of several independent Hessian jäger companies.

The Field Jäger Corps consisted of three Hessian companies under the command of Captain Wrede, Captain Ewald, and Major Prueschenck; an Anspach jäger company commanded by Captain Crammon; a troop of mounted jäger apparently commanded by Captain Lorey; and a 30-man detachment of Hessian grenadiers assigned as infantry support for two three-pounder cannons. Composed of infantry, cavalry, and artillery, the Jäger Corps was a true combined arms unit. At the time of its formation in June of 1777, the corps consisted of approximately 600 men, but by the time the corps embarked for the Chesapeake in early July it apparently mustered 865. Each of the Hessian companies was intended to have 175 men; however, the Anspach company had 105 men described by contemporaries as "tall and handsome fellows." The troops in the first and second companies were trained and experienced soldiers, having fought throughout the campaign of 1776 in New York and New Jersey (Muenchhausen 1974). The third company, however, was composed of men with somewhat more dubious backgrounds, described by Captain Ewald as "deserters and insolent rabble" (Burgoyne 1987:45; Ewald 1979:71).

The jäger were elite troops, trained to fight in open-order formations and to take advantage of cover (Selig 1994). The men composing the Jäger Corps were experienced woodsmen and hunters (literally "jäger") and were regarded as some of the best troops in Howe's army (Knötel et al. 1974:64–65). One British officer wrote in 1776 that the jäger "are a stout, active body of men, armed with rifle-barrel guns, to the use of which they are thoroughly inured, being all without exception chasseurs [hunters] by profession; and as they are commanded by skilful [*sic*], experienced officers, they cannot fail of being very serviceable" (Colonel William Faucett to the Earl of Suffolk, letter, 25 March 1776, quoted in Stephens 1998:51). They were armed with German-manufactured short-barreled rifles with a large bore measuring .65 caliber (Neumann 1967:134; Troiani et al. 1998:48). In his thorough study of flintlock rifles used by the British, DeWitt Bailey notes that, with a .65 caliber bore, the rifles of the Hessian Field Jäger Corps would accept the standard British carbine-bore ball of 0.615 inches, which British rifles also accepted, thus simplifying ammunition supply for the German troops (Bailey 2002:67–68).

British Light Infantry

Two light infantry battalions formed a portion of the vanguard of the British army on the morning of September 3, 1777. Together the battalions mustered approximately 1,130 noncommissioned officers and men (Smith 1976). The light infantry battalions were ad hoc formations created by bringing together the light companies from each of the British line regiments serving in Howe's army. Although both battalions were deployed in the action at Cooch's Bridge, only the 2nd Battalion became engaged with American troops.

At this point in the war, many British officers and men, like the Hessian jäger, were contemptuous of the fighting ability of American soldiers. Lieutenant Purdy of the Pioneers noted that at Cooch's Bridge the Americans "had not forgot[ten] their old Custom," and had been easily routed (Purdy 1777–1778). Captain William Dansey, of the 33rd Light Company, characterized fighting Americans as "Yankee Hunting" to be of no greater concern than "Fox Hunting, at both Diversions a broken bone may be got" (Dansey 1777). Some Hessian officers likened *petite guerre* to hunting as well (Selig and Skaggs 1991:25). The skills of the jäger, particularly in open-order fighting, were graphically described by jäger lieutenant Wangenheim in the summer of 1777 when he wrote home that "we jäger approach the enemy just like on the hunt: we crawl through the bushes on our bellies; upon sighting one, we stalk and shoot him" (Retzer and Londahl-Smidt 1995:3). There is a chilling simplicity in this description, suggesting a measure of self-confidence in the jäger's tactical proficiency and a strong level of contempt and distain for the fighting ability of the Americans.

Like the Hessian Field Jäger Corps, the British light infantry battalions were elite formations composed of the most active, alert, and intelligent officers and men. Recent historical scholarship indicates that a small number of rifles may have been carried by British light infantry companies. Bailey notes there is "inferential documentation" that rifles were carried by dismounted dragoons, some members of the British light infantry companies, and some soldiers in the Guards Division light infantry companies (Bailey 2002:33).

Johann Ewald and *Petite Guerre*

Thirty-three years old in 1777, Captain Johann Ewald (1744–1813) of the Hessian Field Jäger Corps had entered military service at the age of 16, serving in the closing campaigns of the Seven Years' War. He was an extremely competent officer and had been in the service of Frederick II, Landgrave of Hessel-Cassel, since 1760. Ewald studied military engineering in Cassel, in Hesse, and published a volume on tactics in 1774. Hesse had a long tradition of jäger companies, and since 1774 Ewald had served as the captain of the Lieb Jäger Corps (Boatner 1966:356; Selig and Skaggs 1991:18). Ewald was considered by his contemporaries to be a superb master of *petite guerre*, or small war, the sort of small-unit action that was common during the revolution. Ewald considered his experiences in the war to be significant enough for him to publish a book in 1785 on what he called "Partisan Warfare." Entitled *Abhandlung über den Kleinen Krieg* (*Treatise on Partisan Warfare*), the book was widely read by European officers, and its lessons were applied during the Napoleonic Wars in the late eighteenth and early nineteenth century. Ewald's book (referred to here as *Treatise*) drew on previous European studies of irregular warfare but especially on his American Revolution experiences (Selig and Skaggs 1991:9–27).

Ewald took what was already known in Europe about fighting a partisan war or small war and applied his on-the-job experience that he gained in the American war from his rebel opponents. In the eighteenth century in Europe, a new type of infantry was developed to fight small wars. Whereas these units bore distinctive titles in different countries, such as chasseurs, jäger, cazadores, partisans, and irregulars, they were all light infantry. Light infantry was well suited to conducting *petite guerre*, a type of minor operation such as ambushes, raids, reconnaissance, and army security (Birtle 2003:1251; Selig and Skaggs 1991:9–27; Wright 1926:454). Light troops had conducted partisan warfare for decades in Europe prior to the American Revolution, and during the French and Indian War British officers created light infantry companies as a response to irregular warfare in North America (Brumwell 2002; Houlding 1981). At that time, light troops were found to be so successful that British line regiments added a light infantry company to their formations. During the revolution, General Washington also recognized the importance and need for light infantry formations, initially forming ad hoc units such as the corps that fought at Cooch's Bridge. By 1778, the Continental regiment was standardized to include a light infantry company, often combined with other light companies to make elite light battalions, and capable of carrying out the duties ascribed to the light troops.

The officers and men composing light infantry formations were chosen for their strength, abilities, character, and initiative. Essentially, light infantry companies, along with grenadier companies in the British army, were the elite formations. Units of this type were often given the tasks of scouting, covering advances and retreats, and undertaking difficult assignments and assaults. Ewald's writings identified four important characteristics of the light infantryman, qualities that the British, Hessian, French, and American forces strove to develop in their men. Self-reliance, knowledge of terrain and how to utilize it, adaptability to environmental and situational changes, and the ability to innovate and improvise were recommended as the hallmarks of good light troops and by implication as the skills necessary to carry out successful *petite guerre* (Selig and Skaggs 1991:32–37).

Critically to this volume, Ewald's *Treatise* drew on his jäger experiences in North America in how to conduct partisan warfare. As the editors of the

Treatise note in their introductory essay, warfare during the revolution was a "war of small detachments" carried out against and among a civilian population, some of whom were under arms, others who were not. The war in America was an irregular war "fought by regulars against true irregulars" (Selig and Skaggs 1991:19). They conclude that for would-be practitioners of partisan warfare, significant lessons of the revolution included the use of aimed fire, surprise attacks, and ambushes and the use of speed perfected by the rebels in hit-and-run tactics, improvisation, and deception (Selig and Skaggs 1991:20).

Ewald respected his North American adversary, writing at one time that "Americans are the best light infantry-men of the world because every one of them is an experienced hunter and knows all the cunning that is as much part of war as it is of hunting" (Selig and Skaggs 1991:25). He also acknowledged the technical skill and fighting ability of the Americans, writing "what can you do to those small bands who have learned to fight separate, who know how to use any molehill for their protection, and who, when attacked, will run back as fast as they will approach you again, who continually find space to hide? Never have I seen these maneuvers carried out better than by the American militia. . . . If you were forced to retreat through these people you could be certain of having them constantly around you" (Selig and Skaggs 1991:115).

In 1777, at Cooch's Bridge, Ewald was still developing his craft and his approach to fighting American irregulars or light infantry. The battle included at least two and possibly three of the lessons he later wrote about: aimed fire, an initial surprise ambush, and innovation.

Archaeology at Cooch's Bridge

The first-person descriptions of the battle provided principally by the Hessian participants (cf. Burgoyne 1987:47, 1997; Ewald 1979:77–78; Peebles 1998; Retzer and Londahl-Smidt 1998; Robertson 1930; Uhlendorf and Vosper 1935, 1937) record that the American Continentals and militia engaged the British at three separate and distinct battle lines: 1) the point of initial engagement in a woods, possibly an ambush or attack from a concealed position; 2) a second battle line, also in a woods, somewhere between the first line and the final line; and 3) a final battle line on the east side of Christiana Creek and east of Cooch's Bridge. Between these "formal" lines exchanges of fire occurred as battle participants reported that Americans fired, withdrew, reloaded, and fired again as they moved to a new position (Patton 1830; Muenchhausen 1974:28).

Figure 2.2. Location of archaeologically surveyed tracts mentioned in the text. (Courtesy of Wade P. Catts)

The Battle of Cooch's Bridge ranged over an area of about 1,336 ha (5.16 sq mi), roughly from the modern village of Glasgow (formerly Aiken's Tavern) in the south to the intersection of Route 896 and West Chestnut Hill Road to the north (Catts et al. 2014). In recent years several tracts of land contained within these bounds have been archaeologically surveyed: the Gore property, the Sunset Lake tract, and the Cooch tract (Figure 2.2). These studies

Table 2.1. Summary of Lead Shot Recovered from Cooch's Bridge Battlefield

Diameter	Gore property number (%)	Sunset Lake tract number (%)	Cooch tract number (%)	Total number (%)
Up to 0.36 in	0	5 (42)	0	5 (4)
0.37 to 0.60 in	23 (38.3)	7 (58)	23 (49.0)	53 (45)
0.61 to 0.66 in	8 (13.3)	0	11 (23.4)	19 (16)
Over 0.67 in	29 (48.3)	0	13 (27.6)	42 (35)
Total (%)	60 (99.9)	12 (100)	47 (100)	119 (100)

Sources: Catts et al. 2012; Catts et al. 2014; Sivilich 2008.

have recovered over 119 lead shot specimens and other artifacts associated with the battle, lightly spread over the ground (Table 2.1). The areas surveyed through metal detection are not continuous, so accurate density and distribution data is not yet available for the battlefield. However, the types of lead shot and the locations where it has been recovered do provide insights into the fight at Cooch's Bridge and the tactics employed by the opposing forces in this example of *petite guerre*.

Smoothbore firearms of the eighteenth century typically fired a cast, soft lead ball that measured approximately 0.05 to 0.10 inches less than the barrel bore caliber, or size. The difference in size allowed the ball to be more easily loaded down the barrel (as opposed to a breech-loaded weapon), but it also allowed for gas leakage around the circumference during firing (Neumann 1967:14). The difference between the lead shot diameter and the weapon's bore caliber is referred to as windage. Paper cartridges containing a lead bullet (or shot) and a charge of gunpowder were the standard ammunition of the period. Due to the windage, the paper cartridge was necessary to prevent the lead shot from rolling out of the barrel (Peterson 1968:27). From many American Revolutionary War battlefields, archaeologically recovered lead shot with a diameter measuring 0.69 inches is associated with the .75 caliber British Land Pattern musket, sometimes known as Brown Bess. Shot with a diameter measuring 0.64 inches is ascribed to .69 caliber French and/or American muskets (Sivilich 1996:104–105). Large numbers of French Charleville muskets began arriving in the United States in April 1777, so it is likely that some of the American soldiers fighting at Cooch's Bridge were armed with these weapons.

Throughout the battlefield, most of the rifle balls recovered likely come from American riflemen, and a much smaller number are attributable to the jäger companies. In contrast to the standardization seen with the jäger rifle, which fired a ball of 0.615 inches, the bore for American rifles ranged from 0.40 to 0.60 inches, and the length of the barrel averaged 40 inches (Neumann 1967:138). The broader range of ball sizes both smaller and larger than the bullet for the jäger rifle speaks to the presence of rifle-armed militia in Maxwell's corps, men who were specifically drafted from the militia companies to serve in the light corps.

The Sunset Lake and Gore property tracts are in the general area where the battle was initiated. Following the initial ambush of the Hessian advance, Lieutenant Colonel Wurmb distributed his jäger in a battle line apparently anchored on the Newark road (Route 896), using the road as the right flank of the line. The three jäger companies deployed in open order would have occupied about 804 m (.5 mi) of space, the approximate distance from the road to the current fringes of Sunset Lake.

In the Gore property and Sunset Lake fields, the distribution of the lead shot suggests two distinct concentrations. One of these, in the northeastern corner of the Sunset Lake field, includes both dropped (unfired) and impacted shot. This area is roughly situated on the east side of a small knoll, approximately 17 m (56 ft) above sea level. A second concentration of shot is found in the southwestern section of the field on both sides of a slight northwest–southeast trending topographic rise, its highest point 20 m (66 ft) above sea level. This concentration is composed principally of impacted shot. When this data is combined, an interesting interpretation of the archaeological data presents itself. Within the northern Gore property field, 19 specimens of lead shot of all types were recovered by the metal detection survey. Eight shot were attributable to rifles, the balls ranging in size from 0.50 to 0.59 inches. Twelve shot were impacted and concentrated on a knoll or rise located in the middle section of the field. One fired round was recovered in the extreme northeastern corner of the field, on the same northwest–southeast trending rise noted in the Sunset Lake project area. Larger impacted musket balls (two of 0.66 in and one of 0.67 in) were recovered between the two concentrations.

The distance between the shot concentration in the center of the northern Gore property field and the topographic rise in the Sunset Lake project area is approximately 304 m (1,000 ft or 330 yd), within the effective range of an eighteenth-century rifle. These two concentrations of impacted and dropped balls may represent the locations of the initial battle lines. Given that the Battle of Cooch's Bridge was a fluid battle of fire and maneuver, the shot concentrations may be indicative of firing lines for both American and British—or more correctly Hessian—positions, during which time shots were given and received.

The terrain features surrounding Sunset Lake offer additional interpretive information. In September 1777, there was a series of water courses that were apparently being used by the Americans as natural obstacles, and we have already discussed the "impassable morass" or swamp that impeded the flanking movements of the British light infantry battalion. The American formations were apparently using a stream called Five-Mile Run (the precursor to today's Sunset Lake) as the anchor for the left flank of their battle line. They also used the elevated northwest–southeast trending rises as battle positions as evidenced by the scatters of shot in both the northern Gore property field and the Sunset Lake tract. The smaller, unnamed streams that enter Five-Mile Run west of Sunset Lake, one of which now forms the western arm of the lake, functioned as key natural terrain obstacles that fronted the American positions. The period descriptions of the battle describe the initial American line as covered in woods so it seems that the fields on the Gore property and the Sunset Lake tract were at the time of the battle wooded or tree covered. The number of specimens of impacted lead shot recovered from these fields also tends to support this conclusion.

When compared to the metal detection findings from other areas surveyed on the battlefield, the Sunset Lake area contained a comparatively large number of buckshot. Previous studies of the adjacent Gore property recovered four specimens of modern buckshot but none that were battle related (Sivilich 2008:8). In addition, the battlefield survey of the lands surrounding Cooch's Bridge yielded no buckshot (Catts et al. 2014). Thus, the buckshot on the Sunset Lake property could be related to the battle, or, as possible, post-battle hunting. However, based on their level of patination, size, method of construction, and distribution, at least five of the buckshot, along with the recovered musket and rifle balls, are most likely associated with the battle.

Although not definitive, the presence of early buckshot may identify American musket men. The use of buck and ball cartridges by Continental army soldiers was common beginning in the early years of the war, and in June 1776 General Washington recommended that for initial volleys, muskets be loaded with one musket ball and from four to eight buckshot, depending on the type of musket (Washington 1776). Whereas the use of a buck and ball load was recognized as common practice in the Continental army, it was a month after the Battle of Cooch's Bridge, on October 6, 1777, that Washington made the practice standard for his troops, ordering that "buckshot shall be put into all cartridges which shall hereafter be made" (Washington 1777b). Archaeological examples of buck and ball cartridges were recovered on the Camden battlefield, where two complete loads of one musket ball (.69 caliber) and three buckshot were found in an unplowed context. Metal detecting in the surrounding field recovered a cache of nine musket balls (.69 caliber) with 27 buckshot (Legg et al. 2005:104). A wide range of buckshot sizes has been recovered from other Revolutionary War battlefields, including Monmouth, Camden, and Waxhaws (Sivilich 2009; Legg et al. 2005:102–104; Smith et al. 2009:70; Butler 2011).

At the Cooch tract, archaeological survey was undertaken in the south, east, and west cornfields. The retrieval of shot from the south cornfield, located a short distance south of the bridge, reveals battle-related action. Fighting in the south cornfield may represent action that occurred between the second and third battle lines when the American soldiers were firing, retreating, reloading, and firing. The lead balls recovered in the south cornfield represent rifle and musket balls, both impacted and dropped. Although not conclusive, two clusters of shot were identified: one at the extreme northern end of the field composed chiefly of impacted shot ranging from 0.40 to 0.49 inches (rifle shot), and a second cluster along the eastern edge of the field, approximately 137–152.4 m (450–500 ft) away from the first cluster.

The second cluster contains 0.40 to 0.49 inches impacted shot, 0.50 to 0.59 inches dropped and impacted shot, and 0.60 to 0.66 inches dropped and impacted shot. This cluster seems to represent a broader range of weapons, including rifles and muskets. Taken together, the two clusters suggest firing lines at several hundred feet from each other. Because no metal detection was undertaken in the wooded area north of the first cluster, it is difficult to know whether this cluster represents a target line from an area closer to Cooch's Bridge. Additionally, considering the fluid character of the battle on September 3, it is likely that target lines and firing lines were essentially interchangeable; firing lines were overrun and, once occupied by opposing troops, became target lines (Figure 2.3).

The east cornfield survey area is situated east of Christiana Creek. The recovery of dropped (n = 4) and impacted (n = 9) musket and rifle balls from the east cornfield suggests that the final phase of the battle moved beyond the bridge and continued into the fields east of the creek. The lead balls found here are almost exclusively large diameter (.66 in or greater) or rifle sized (0.50–0.59 in), perhaps representing the firing of British light troops and Hessian Field Jäger into American light infantry prior to their retreat to the east along the King's Highway toward the village of Christiana Bridge.

Other classes of artifacts that were recovered and have military-related potential include a wide variety of buttons (n = 21) and a broad distribution of period coins. In addition, whole and partial buckles and several iron kettle fragments were recovered, most coming from the immediate vicinity of the Cooch house and from the east cornfield. These artifacts are attributable to the American prebattle camp around the Cooch house and mill, or to the postbattle five-day British camp. No definitive marked regimental buttons

Figure 2.3. Location of surveyed fields on the Cooch tract. (Courtesy of Wade P. Catts)

from American or British formations have been found on the battlefield, but at least four buttons recovered during the Cooch tract survey are tentatively attributed to Hessian units, probably the jäger companies. These buttons are plain, so they could equally be attributed to nonmilitary sources, although all four of these buttons display the characteristics of buttons associated with Hessian uniforms (Powell 1996; Troiani 2001:76). Three buttons recovered from the east cornfield conform to Powell's type 1, subtype 2 classification.

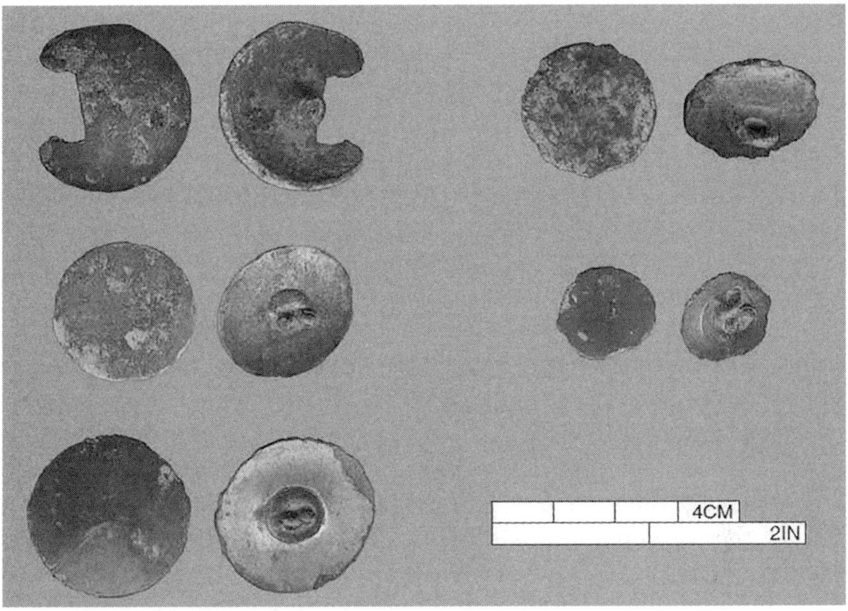

Figure 2.4. Possible Hessian military buttons recovered on the Cooch tract. (Courtesy of Wade P. Catts)

Two measure 22 mm (.86 in), and the third is larger in size (26 mm [1.20 in]) and is type 1, subtype 1A. The single possible Hessian button (16 mm [.62 in]) recovered from the west cornfield is also type 1, subtype 2. Several large, plain buttons, like those found in the east and west cornfields, were recovered from the Gore property fields in 2008 and more recently from the archaeological survey of the Red Bank battlefield in New Jersey, an engagement that was fought exclusively by a Hessian brigade against American forces (Figure 2.4) (Catts et al. 2017; Sivilich 2008).

Conclusion

In comparison to other battlefields of the American Revolution where in-depth archaeological investigations or studies have been completed, such as Monmouth, Bennington, Hubbardton, Red Bank, Camden, Cowpens, or Eutaw Springs, the number of recovered battle-related artifacts from Cooch's Bridge is understandably low (cf. Babits 1998; Catts et al. 2017; Kenny and Crock 2010; Legg et al. 2005; Selig et al. 2016; Sivilich 2009; Smith et al. 2009). This is not unusual, because these other battles were much larger engagements in terms of duration and numbers engaged. At the same time, the number of battle-related artifacts recovered from the larger Cooch's Bridge battlefield is relatively high when compared with other Philadelphia campaign fields of conflict, such as Princeton, Brandywine, Paoli, and Germantown, indicating the relatively excellent state of preservation of that battle landscape.

The study of battlefields requires a unique combination of military history, archaeology, and terrain or landscape analysis (Foard and Morris 2012). In the case of the Battle of Cooch's Bridge, the combination of these three elements has resulted in a more detailed understanding of the military action of September 3, 1777. The archaeological evidence of the battle provides physical evidence of how the battle was fought and helps to underscore the opponents' tactics. Although portions of the battlefield remain to be archaeologically surveyed, the archaeological evidence so far discovered allows for several interpretations. The scale of the engagement is representative of many of the small-unit actions fought during the war. The degree of battle intensity is not that observed at major battles, such as Camden, Monmouth, or Red Bank. The archaeological signature of the Battle of Cooch's Bridge is extensive rather than intensive. It covers a large geographic area and is defined by concentrations of battle-related artifacts separated by areas of sparse artifact presence. The archaeology reveals a battle of set firing lines, established for fire control of the infantry units involved. At the same time, the artifact distribution suggests that the battle lines were extremely fluid to allow for advance and retreat. The recovered lead shot includes musket balls, rifle balls, and buckshot. A relatively narrow range of rifle ball diameters represents the Hessian jäger rifles. The much larger range of diameters present on the field represents the weapons of the British light infantry and the various calibers used by American Continental and militia formations at this period of the war. The American firing lines are also locations of hand-to-hand combat, well described in the first-person accounts. These locations are marked by dropped and impacted shot but should also include other military "detritus," such as broken buckles, lost buttons, cartridge box plates, and discarded or abandoned items. Further archaeological study could focus on these battle line locations, as it is likely that artifacts associated with the combatants are extant, thus providing additional archaeological information about the battle that is not available in the documentary record. The battle line locations, where casualties were taken by both Hessians and Americans, may also be the locations were battlefield burials occurred.

The action at Cooch's Bridge is an example of an asymmetric engagement fought between two contrasting military formations. The Hessian Field Jäger Corps was a well-trained elite light infantry unit composed of infantry, dragoons, and artillery, supported by equally elite British light infantry and British grenadier battalions. In contrast, the American light infantry corps was a newly formed, ad hoc unit combining musket- and rifle-armed soldiers, only some with previous combat experience. Although well led by seasoned officers, the unit lacked artillery and, more importantly, the cohesion that only long experience and service could provide. In the late summer of 1777, the American light infantry corps was an expedient for General Washington and an experiment in developing elite infantry. Within the year, light infantry companies would be the norm as part of every Continental regiment. The light troops on both sides, including various mixed- or combined-arms units, sometimes called partisans, legions, chasseurs, or jäger, would increasingly become the military units most often called on to conduct *petite guerre*. Cooch's Bridge was an opportunity for Hessians, British, and Americans to learn their craft. Most importantly, Johann Ewald used the action at Cooch's Bridge as an example in his *Treatise*, and American soldiers came away with valuable lessons in the art of *petite guerre*.

Acknowledgments

Many colleagues have assisted me in my work at Cooch's Bridge. Foremost among these are the late Edward W. Cooch Jr., Richard Cooch, Barbara Cooch, Anne Cooch Doran, and their families, who have tirelessly preserved the battlefield and allowed archaeological survey on their lands. I would also like to thank Daniel Sivilich, Eric Sivilich, and the metal detector operators with Battlefield Restoration & Archaeological Volunteer Organization (BRAVO). The members of BRAVO, along with volunteers from the First State Search and Recovery Club, have steadfastly assisted me in the various battlefield surveys at the site. Funding for some of the work has been provided by the Delaware Division of Historical and Cultural Affairs (Timothy Slavin, Craig Lukezic, and Lynn Riley) and the Department of Natural Resources and Environmental Control (Cherie Clark). My Cooch's Bridge research has been generously supported by a grant from the Delaware Heritage Commission. Others who have contributed to the work or to my thinking and interpretations of the battle include, in alphabetical order, Charles Fithian, Clarence R. Geier, Juliette Gerhardt, Elisabeth LaVigne, Peter Leach, Timothy Mancl, David Orr, Robert Selig, Steven D. Smith, and Scott Stephenson. I also thank William Lees for his insightful review of the chapter.

3
The Battle of Williamson's Plantation

Huck's Defeat and the Asymmetric Partisan War in the South Carolina Backcountry

Michael C. Scoggins and Steven D. Smith

The American Revolution is often pictured in the popular media as a conventional war between the American Continental army and the British regular army, replete with large land battles employing conventional eighteenth-century linear military tactics. There is, however, another aspect to the war that was at least as significant to America's ultimate independence as the large conventional battles, and even more typical of the actual conflict, especially in the southern campaigns of the revolution. Throughout the Carolinas and Georgia, particularly along the Appalachian and Floridian frontiers, the American Revolution was more often a much different type of war. Here the battles generally involved smaller groups of combatants, most of whom were irregulars, engaged in a partisan war that enveloped men, women, and children of all races and social classes. Some involved Patriot militias (whom the British universally described as "rebels") fighting other militia groups that were loyal to the British Crown (Loyalists), and some involved frontier militias fighting against Native Americans in a war that was more about territorial expansion and self-preservation than independence. Still others involved American revolutionary or "Patriot" militias fighting British Crown and provincial troops. This latter scenario was illustrated by the action at Williamson's Plantation, South Carolina, on July 12, 1780, with the addition of Loyalists in the mix.

Partisan Warfare in the South

From 1775 through early 1780, the campaigns in the southern theater of the American Revolution involved numerous engagements representing both conventional (symmetric) and unconventional (asymmetric) warfare, but from the American point of view the war was still waged under the direction of state and Continental authorities. In South Carolina, the militia, state troops, and Continental regiments all answered to the state government, and all major campaigns and troop movements were organized and directed by the governor and General Assembly in conjunction with the Continental Congress. The conflict during this phase of the war included numerous "symmetric" battles between American and British regular troops, often supported by Patriot and Loyalist militias, as well as battles that pitted Patriot and Loyalist militias against each other without the support of any regular troops (Buchanan 1997; McCrady 1902; Scoggins 2005).

There were, however, numerous examples of asymmetric warfare during the early phase of the war as well. These included several campaigns launched by the southern states against the British-allied Cherokee and Creek in residence along their frontiers. The largest of these took place in the summer and fall of 1776, when the state troops and militias of the two Carolinas and Virginia undertook a major, joint punitive campaign that effectively destroyed the Cherokees' ability to make war for years to come (Dean 2012; Dennis 2017; Scoggins 2005). This warfare was simply a southern extension of a much greater struggle for control of the Appalachian frontier throughout the 1770s and 1780s that extended from New York and Pennsylvania through the Shenandoah Valley of Virginia and into the Carolinas and northern Georgia. (See chapter 1 for an example of this conflict in the Greenbrier Valley of present-day West Virginia, which was part and parcel of this larger transatlantic conflict.)

The military situation in the South changed drastically in May 1780, when the British army and navy captured Charleston and an entire 5,000-man army stationed there—an army that included all of South Carolina's Continental troops, a substantial number of North Carolina and Virginia Continentals, and a large body of state militia. The British commander, Sir Henry Clinton, and his subordinate, Lieutenant General Charles, Earl Cornwallis, confidently expected that the surrender of Charleston would put an end to all resistance in South Carolina. This, in turn, would allow the Crown forces to move rapidly to subdue North Carolina and then advance into Virginia. If successful, this would cut off the valuable southern colonies from congressional control and force the fledgling United States to sue for peace (Buchanan 1997; Chacon and Scoggins 2014; Edgar 2001; McCrady 1902; Scoggins 2005; Tarleton 1787).

This expectation on the part of the British high command soon proved to be premature. What Clinton and Cornwallis failed to anticipate was that the loss of Charleston and the dissolution of the revolutionary state government would not end the opposition to British control. In the interior, or "backcountry," of South Carolina, there were still large numbers of rebels who were unwilling to submit to British authority, and these rebels were quite capable of fighting a partisan war without the support of the state or

Continental governments (Tarleton 1787; see also Smith, chapter 4 in this volume). This was especially true among the so-called Scots-Irish Presbyterians who had immigrated to the Carolina backcountry from Lowland Scotland and the province of Ulster in Ireland. These militant, stubbornly Calvinistic Protestants had a long history of conflict with the British Crown and the Church of England, and they brought their dislike of British imperialism with them to the Carolinas (Buchanan 1997; Chacon and Scoggins 2014; Edgar 2001; McCrady 1902; Scoggins 2005). There were also substantial numbers of Loyalists in the backcountry, who were, at least nominally, willing to support the Crown troops in their efforts to pacify South Carolina.

As British troops began occupying the western settlements and establishing fortified positions along the northern frontier, the war rapidly took on new characteristics that the British had not foreseen. The Loyalists in South Carolina had been suppressed by the revolutionary state government since the beginning of the war. With the British now in control of the state, the Loyalists began organizing their own militia regiments, offering their services to the Crown, and exacting vengeance on their rebellious neighbors. Rather than submit to the British, however, the rebels in the backcountry, particularly the Scots-Irish Presbyterians, organized their own militia regiments and took the offensive against the British and their Loyalist allies.

The war then became both an all-encompassing civil war, pitting loyal Carolinians against rebel Carolinians, and a full-blown partisan war, as bands of Patriot troops struggled against British regular and provincial soldiers for control of the interior. The "civil war" aspects of the revolution in the Carolinas had been a factor from the beginning of the war, as the rebels ("Whigs") and Loyalists ("Tories") had been fighting each other for years. Now, however, with the British nominally in control after the fall of Charleston, the violence between the two factions escalated drastically (Buchanan 1997; Edgar 2001; McCrady 1902; Scoggins 2005; Tarleton 1787; Weigley 1970).

Similarly, the asymmetric aspects of the war had been widespread during the earlier years of the revolution, as Loyalist partisans tried, unsuccessfully, to undermine the control of the revolutionary government in Charleston. Driven from their homes, many of these Loyalists fled to the swamps of British-controlled East Florida, which they used as staging areas for raids into southern Georgia and western South Carolina. After May 1780 the situation was reversed; the Whigs were now the partisans, fighting desperately to undermine the success of Clinton's "Southern strategy" and an army of well-trained, professional British soldiers (Buchanan 1997; Edgar 2001; Scoggins 2005; Weigley 1970).

The partisan nature of the war in the Carolina backcountry was obvious to commanders of both armies even as it was happening. The infamous British officer Lieutenant Colonel Banastre Tarleton, who commanded a provincial unit of cavalry and infantry called the British Legion, frequently described the Whig militia commanders in South Carolina as "active partisans" (Tarleton 1787). Tarleton's American counterpart, Lieutenant Colonel Henry "Light Horse Harry" Lee of the Continental army, referred to his battalion of cavalry and infantry as the "Partizan Legion" (Lee 1998 [1812]). The North Carolina cavalry commander Major William Richardson Davie described his troops as a "partisan corps" (Davie 1976). Subsequent historians during the nineteenth and twentieth centuries were also well aware of the partisan aspects of this asymmetric conflict. As Edward McCrady, a former Confederate field officer, noted in his detailed study of the revolution in South Carolina, "Since the fall of Charleston there had been really no militia in the State, though the partisan bands were usually so called; for a militia . . . implies the existence of a state government under which the citizens are enrolled and required to do duty. But since [South Carolina Governor John Rutledge's] departure from the State there had been no government except that of the British authorities under the protection of the Royal army. . . . These bands were thus purely volunteers fighting from Patriotism only, without pay or reward" (McCrady 1902:511–512).

Williamson's Plantation/Huck's Defeat

One of the earliest and most significant of the partisan battles along the northern South Carolina frontier is known as the Battle of Williamson's Plantation or, more simply, as "Huck's Defeat." This engagement between Whig militia and British provincial soldiers took place two months after the fall of Charleston and became an archetype for much of the subsequent partisan warfare in the Carolina backcountry (Buchanan 1997; Edgar 2001; McCrady 1902; Scoggins 2005).

Following the fall of Charleston in May, Clinton turned command of his southern army over to Cornwallis and returned to New York. Cornwallis quickly occupied the northern frontier towns of Camden, Ninety Six, and Long Bluff and garrisoned them with detachments of British regulars, provincials, and Loyalist militiamen. Cornwallis also established several fortified outposts further inland including Rocky Mount, a high elevation overlooking the area where Rocky Creek enters the Catawba River, in the heart of the South Carolina up-country. Rocky Mount was commanded by an experienced British officer, Lieutenant Colonel George Turnbull, and was garrisoned by approximately 150 troops, composed of both British provincial soldiers and Loyalist militiamen (Edgar 2001; McCrady 1902; Scoggins 2005; Tarleton 1787).

The provincials at Rocky Mount included a company of Lieutenant Colonel Turnbull's light infantry regiment, the New York Volunteers, commanded by Lieutenant William Adamson of New York. Also at Rocky Mount was a troop of British Legion dragoons or light cavalry under Captain Christian Huck of Philadelphia. These provincials were experienced veterans of the war in the northern states as well as in the battles of Savannah and Charleston. Huck's troop had fought in the infamous Battle of Waxhaws on May 29, 1780, where British Legion dragoons and infantry commanded by Lieutenant Colonel Tarleton reportedly massacred a battalion of American Continental soldiers after they had surrendered.

The Rocky Mount Loyalist militia regiment was organized into two battalions commanded by Colonel Matthew Floyd from present-day York County, and Colonel James Ferguson from present-day Chester County. Unlike the provincials, the Tory militia had little military experience, and their commander at Rocky Mount, Colonel Turnbull, was not very impressed by their appearance or their performance (Scoggins 2005).

In June 1780, Turnbull dispatched Huck's dragoons and a company of Loyalist militia to destroy Whig militia camps at Fishing Creek Presbyterian Church in Chester County and William Hill's ironworks in York County, both of which were centers for rebel activity (Figure 3.1). Huck destroyed both the Fishing Creek Meeting House and the Hill's Ironworks factory complex. He then conducted extensive foraging operations throughout the area, confiscating wheat, corn, cattle, horses, and slaves from rebel plantations, all of which he took back to Rocky Mount. He also became notorious for his heavy-handed treatment of the rebels, which included burning Whig plantations and hanging Whig militia leaders (Moore 1859; Scoggins 2005; Winn 1812).

Following the loss of their two field bases, the Whigs from the area between the Broad and Catawba Rivers, most of whom were also from the present-day counties of York and Chester, moved to the east side of the Catawba River (see Figure 3.1). Here they established a camp at the Old Nation Ford (named for the Catawba Indian Nation) and began organizing a partisan militia brigade under the command of Thomas Sumter, then a colonel. Sumter was the former commandant of the 6th South Carolina Continental Regiment and would later be elected as brigadier general of militia (Craig 1854; Hill 1921; Moore 1859; Scoggins 2005; Winn 1812). Sumter and many of his field officers, including men like William Bratton from present-day York County and Edward Lacey from present-day Chester County, were veterans of the French and Indian War along the Pennsylvania and Virginia frontiers and had seen extensive service during the first five years of the revolution in the Carolinas and Georgia. These men had witnessed the disastrous de-

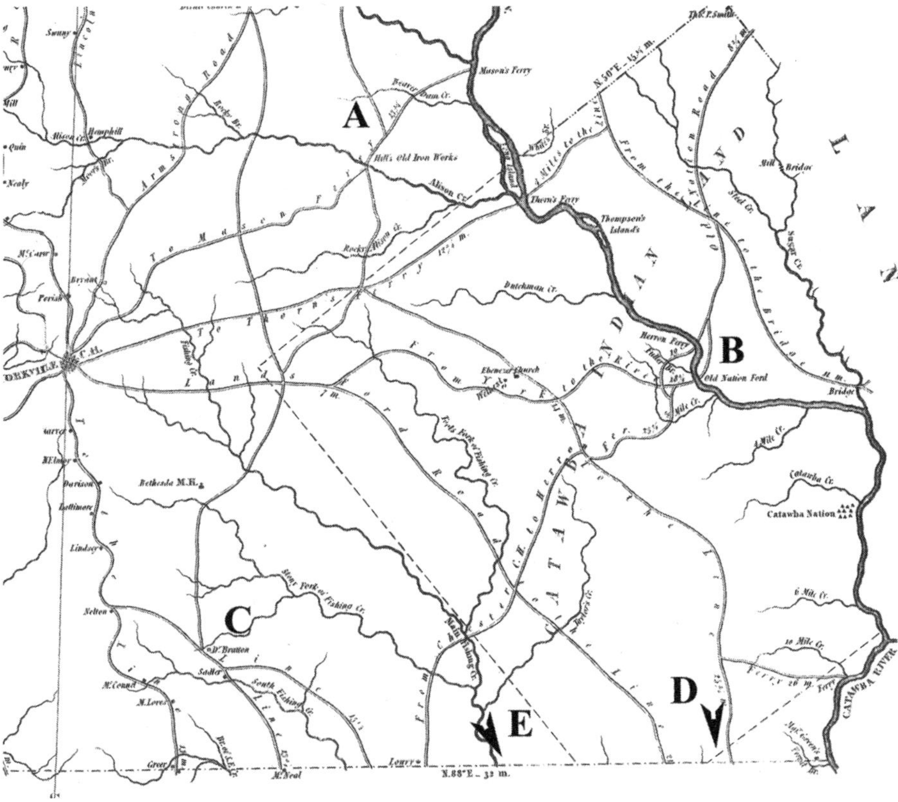

Figure 3.1. Map depicting general locations of camps and battlegrounds in York District of South Carolina. A = Hill's Ironworks, B = Nation Ford camp, C = Bratton's and Williamson's plantations, D = Rocky Mount, 32 km (20 mi) downstream, E = Walker's Mill, 4.8 km (3 mi) downstream. (Adapted from Robert Mills's 1825 *Atlas of the State of South Carolina*)

feats of British troops and colonial militias on the Appalachian frontier during the French and Indian War. They had also experienced, firsthand, the unfortunate results that occurred when Continental generals tried to fight the British army on its own terms in the earlier campaigns of the revolution. Accordingly, they eschewed eighteenth-century military conventions and employed the hit-and-run tactics that Native Americans had utilized so successfully against both British and American troops along the frontier. These tactics included ambushes of isolated enemy detachments, especially in the early morning hours; capturing enemy supply convoys; and most of all, employing their horses to conduct a fast-moving, mobile type of warfare among the hills and valleys of the Carolina up-country (Scoggins 2005).

These same tactics had served Native Americans well against their neighbors and in the asymmetric war against the Europeans since the latter's arrival on the continent and up through the late eighteenth century (see Sivilich, chapter 5, and Haecker, chapter 8).

In early July 1780, Turnbull received intelligence that many of the local Whig militiamen, including Captain John McClure and Colonel William Bratton of the Fishing Creek communities in Chester and York Counties, had returned home to check on their wheat harvest and to enlist additional recruits for Sumter's Brigade. Turnbull gave Huck instructions to apprehend McClure, Bratton, and other rebel leaders, and to engage and disperse the Whig militia wherever he might find them. On the evening of July 10, Huck set out from Rocky Mount with 35 British Legion dragoons, 20 New York Volunteer mounted infantrymen, and 60 Tory militia rangers. Over the course of the following day, Huck's battalion slowly worked its way north through Chester County up the old Rocky Mount Road into York County, making numerous stops along the way to arrest rebel militiamen and confiscate supplies for the Rocky Mount garrison (Craig 1854; Edgar 2001; Moore 1859; Scoggins 2005).

Thanks to intelligence from several local residents, Sumter's men soon learned that Huck was once again on patrol, and they quickly made plans to intercept him. Throughout the day on July 11, the Whig officers dispatched express riders to round up volunteers from all over present-day York and Chester Counties. Believing that Huck was camped at Phillip Walker's gristmill in Chester County (where the town of Lando, South Carolina, is located today), the Whigs set off from the Nation Ford camp late on the evening of July 11 with about 200 men (see Figure 3.1). Their plan was typical of the frontier tactics that they had learned during the French and Indian War: advance on the enemy under cover of darkness and surprise it in a dawn attack. The primary Whig militia commanders were Colonel Bratton, Colonel Andrew Neel, and Lieutenant Colonel William Hill from York County; Colonel Edward Lacey and Captain McClure from Chester County; and Captain Richard Winn from Fairfield County (Craig 1854; Hill 1921; Scoggins 2005; Winn 1812).

Meanwhile, Huck's force continued north into York County, headed for William Bratton's plantation (see Figure 3.1). The New York Volunteers and the Tory militia arrived at the Bratton home late on the afternoon of July 11. Colonel Bratton's wife, Martha, and some other family members had just come in from the fields after reaping wheat all day. A Tory militiaman demanded to know the whereabouts of Martha's husband and threatened her with a reaping hook when she refused to answer. Lieutenant Adamson of the New York Volunteers quickly came to Martha's rescue and protected her from the belligerent Tory. Huck and his British Legion dragoons then arrived on the scene. After an unproductive and frustrating interview with Martha, during which she refused to cooperate in any way, Huck locked Martha and her children in the attic of her house. He then moved his troops to the neighboring home of James Williamson, who had a large field of oats that Huck wanted for his horses (Moore 1859; Scoggins 2005).

Oral history and maps provided by three of William Bratton's grandsons (John, Rufus, and Napoleon Bratton), as well as by James Williamson's nephew John Starr Moore, all agree that the Williamson plantation was located approximately 275–365 m (300–400 yd) southeast of Bratton's house, down a long sloping hill toward what was then known as Becky's Branch, a tributary of the South Fork of Fishing Creek (Figure 3.2). The Williamson plantation included a two-story log house, a corncrib, and a stable or barn, all located in a hollow and flanked by several spring branches. The area around the house was cleared. To its south was a fenced-in lane, "Williamson's Lane," that ran southeast to northwest along a ravine back up the hill toward Bratton's house. There the lane intersected the Armstrong Ford Road, a heavily traveled colonial highway that ran north to south along the west side of Bratton's house (roughly congruent to modern Brattonsville Road). To the south of the lane was a cultivated field where oats were probably planted. On the west and/or north sides of the house were one or more orchards planted with apple, peach, and plum trees (Scoggins 2005; Scoggins et al. 2011).

When the various Whig companies arrived at Walker's Mill on the night of July 11, they found that Huck had moved north to the Bratton plantation. During a forced march from Walker's Mill into York County, about 50 Whig militiamen dropped out and either returned to camp or went home. The remaining Whigs, now about 140 in number, picked up Huck's trail on the Rocky Mount Road and then crossed over to the Armstrong Ford Road, a few miles south of the present-day York-Chester County line. They arrived near Bratton's plantation about 3:00 a.m. Approximately 1.6 to 2.4 km (1 to 1.5 mi) south of Bratton's, the Whigs dismounted, secured their horses off the main road, and proceeded on foot. Believing that the Crown forces were camped at the Bratton homestead, the Whig commanders dispatched an advance party of about 25 mounted men under a Captain Read to swing around the west side of the home, outflank the enemy, and cut off its avenue of retreat. The main force was to attack up the main road from the south (Hill 1921; Moore 1859; Scoggins 2005; Winn 1812).

On reaching the intersection of the Rocky Mount Road and the Armstrong Ford Road, approximately 457 m (500 yd) south of Bratton's house, the Whigs received new intelligence from local residents that Huck was

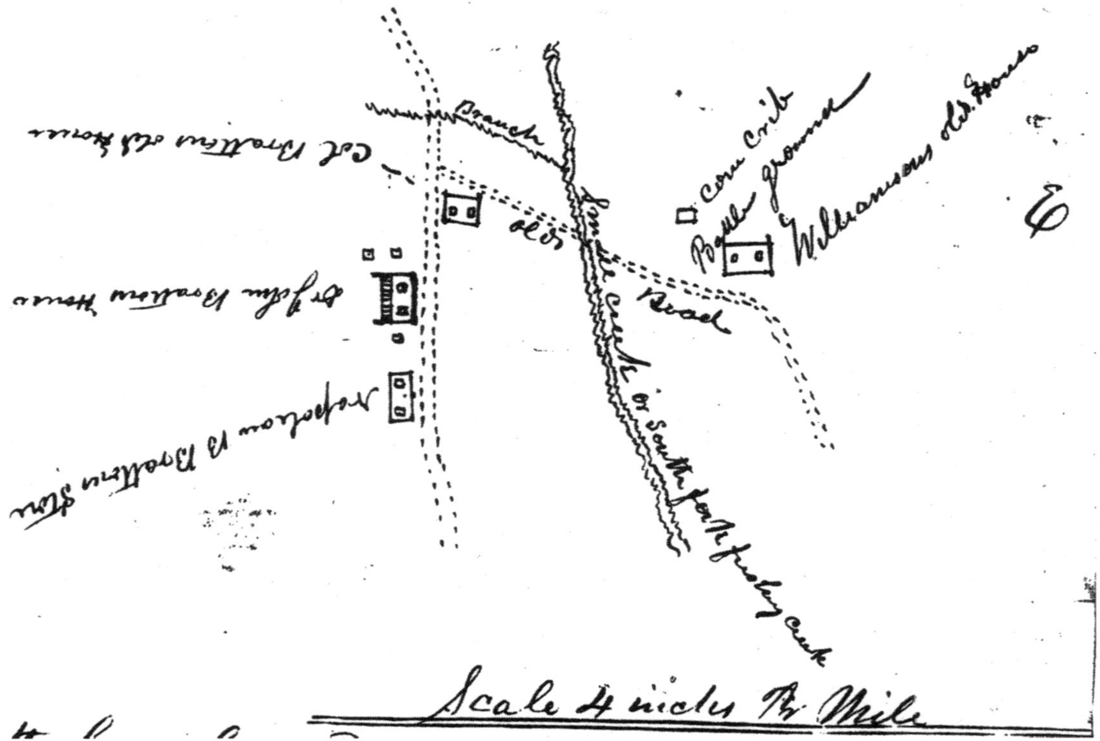

Figure 3.2. Map of Bratton's and Williamson's plantations drawn by Daniel G. Stinson, March 1876. Sumter Papers, Draper MSS, 15V278, Image No. WHi-27323. (Courtesy of the Wisconsin Historical Society, Madison)

actually camped at James Williamson's plantation (Moore 1859; Scoggins 2005; Winn 1812). Informants thoroughly briefed the Whigs on the layout of the enemy camp: the Tory militia was positioned in an old field about 91 m (100 yd) south of Williamson's house, whereas the New York Volunteers were camped in the fenced-in lane. The British Legion dragoons were positioned around the Williamson house, with Huck and the other officers billeted inside (Hill 1921; Scoggins 2005; Winn 1812).

Of the Whig soldiers who can be documented as participating in Huck's Defeat, at least half lived in the nearby neighborhoods around Fishing Creek Presbyterian Church and Bethesda Presbyterian Church. These men were very familiar with the layout and terrain of the land, and once they received the new intelligence of the enemy's disposition, they knew exactly where they would find the British and Loyalist campsites (Moore 1859; Scoggins 2005).

At this point the Whigs decided to divide their force again and attack the British from the east and west ends of Williamson's Lane. One group, consisting primarily of men from present-day York County, commanded by Bratton and Neel, turned northeast from the road and marched diagonally across the southern fields of Bratton's and Williamson's plantations to attack the Loyalist militia camp from the southwest. The second group, composed primarily of men from present-day Chester County and commanded by Lacey and McClure, circled widely around the south side of the British camps toward the east end of Williamson's Lane. These men had to traverse some very difficult terrain, including creek swamps and wooded areas. Consequently, they were delayed getting into their position (Collins 1859; Craig 1854; Hill 1921; Scoggins 2005; Winn 1812).

Just as the sun began to rise, about 5:00 a.m., the Whigs under Bratton and Neel were spotted by a Loyalist sentinel. Having been spotted, they commenced their attack by prearranged signal. The Loyalist militiamen, who were eating breakfast and preparing to break camp, were caught completely by surprise. The senior Loyalist militia officer, Colonel Floyd, mounted his horse and fled the battlefield, along with many of his men (Hill 1921; Moore 1859; Scoggins 2005; Winn 1812). The other Loyalist militia commander, Colonel Ferguson, stood his ground and tried to rally his men, only to be shot down at almost point-blank range by the vengeful Whigs (Craig 1854; Moore 1859). Many of the Tory militiamen were killed or wounded by the Whig riflemen, and many more surrendered. Others abandoned their horses and weapons on the field and escaped on foot to the surrounding woods. Some of the escapees died from their wounds after fleeing into the woods, their bodies being discovered during the following week (Hill 1921; Scoggins 2005; Winn 1812).

Following the rout of the Loyalist militia, the Whigs rapidly advanced on the New York Volunteers, who were camped in Williamson's Lane. The Volunteers were Provincial infantry well trained in the bayonet charge, but they were hemmed in by the fences running along both sides of the lane and could not maneuver effectively. They quickly began taking casualties from the Whigs' rifle fire and, seeing the hopelessness of their position, grounded their arms and surrendered. The New York Volunteers' commanding officer, Lieutenant Adamson, attempted to escape up Williamson's Lane and was mortally wounded when he fell from his horse and was impaled through the chest by a pine sapling. The Whigs under Bratton and Neel then turned their attention on the British Legion dragoons who were camped around Williamson's log house (Collins 1859; Craig 1854; Hill 1921; Moore 1859; Scoggins 2005; Winn 1812).

When the Whigs began their attack, Huck rushed out from the Williamson house and began shouting orders to his men. The British troopers mounted their horses and formed up, intending to charge the enemy soldiers as they had done so successfully in previous battles. But the Whigs remained behind the peach trees and fence rails, taking careful aim with their rifles as the dragoons waved their swords and shouted curses at them (Collins 1859; Craig 1854; Moore 1859; Scoggins 2005). After about a half dozen of Huck's men were killed or wounded, Huck and several of his officers tried to break out of the trap and spurred their horses up Williamson's Lane toward Bratton's house. As Huck galloped away from the battle, a group of Whig riflemen took aim and fired. One of them, a Fishing Creek militiaman named John Carroll, loaded two balls in his weapon before firing. Huck was hit and fell from his saddle to the ground, dead. When his body was examined, two bullet holes were found in the back of his head, about 12.7 mm (.5 in) apart, and Carroll was given credit for firing the shot that killed Captain Huck. With their captain dead, and the second-in-command rapidly fleeing the battlefield, the remaining dragoons surrendered (Collins 1859; Craig 1854; Scoggins 2005; Winn 1812).

The Whigs under Lacey and McClure were delayed reaching their positions and missed out on most of the battle. After the battle ended, McClure freed a group of Whig prisoners that Huck had locked in Williamson's corncrib. Among these prisoners were McClure's youngest brother, James, and one of William Bratton's older brothers, Robert. Captain McClure and some of his men then mounted up and took off after the Loyalists who had escaped on horseback, pursuing them almost all the way back to Rocky Mount (Hill 1921; Moore 1859; Scoggins 2005).

Typical of successful asymmetric engagements, the fighting was over in fifteen minutes or fewer, and the casualties were one sided. The total number of provincial and Loyalist casualties was approximately 30 killed and 50 wounded, and a large number were taken prisoner. The only confirmed Whig casualty was a man from Chester County named Campbell. After the battle ended, Campbell was escorting a Tory prisoner up to the Williamson house at gunpoint when the Tory pulled a pistol from inside his coat and shot him at point-blank range. Campbell was killed instantly, and the Tory made his escape (Scoggins 2005).

The day after the battle, the Whigs loaded up the captured and wounded Loyalists in wagons and sent them back to Rocky Mount as prisoners on parole. The provincial and Loyalist casualties were buried in shallow, unmarked graves somewhere on the lower end of the Williamson property. The Whigs' victory was complete. Their destruction of Huck's force at the Battle of Williamson's Plantation on July 12, 1780, helped revive the morale of the people in South Carolina just when the situation seemed darkest. The battle served as a rallying point for the backcountry Whigs and brought several hundred fresh recruits into General Sumter's partisan militia brigade. The battle also set into motion a series of significant events that would soon lead to even larger Patriot victories at Musgrove's Mill in August 1780, King's Mountain in October 1780, Blackstock's Plantation in November 1780, and Cowpens in January 1781, and ultimately to the British surrender at Yorktown in October 1781 (Buchanan 1997; Edgar 2001; Hill 1921; McCrady 1902; Moore 1859; Scoggins 2005).

Archaeology of Williamson's Plantation

The site of the Battle of Williamson's Plantation, or Huck's Defeat, as it soon became known throughout the Carolinas, remained an attraction to local visitors for many years. In 1787 the Williamsons sold much of their property, including the battle site, to Colonel Bratton, who dismantled the Williamson house and turned the land into part of his expanding plantation, subsequently known as Brattonsville. Colonel Bratton's son, John Simpson Bratton, hosted a large celebration at Brattonsville on the battle's anniversary in 1839. This celebration included a visit to the battlefield by several surviving veterans who pointed out the scenes of the battle. The Loyalist graves were still visible and were frequently visited by residents as late as the 1870s (Scoggins et al. 2011:27). By 1915, however, the Bratton family members had all moved away from Brattonsville and turned the land over to tenants, and the exact site of the battle was gradually forgotten. Brattonsville, including the Huck's Defeat battlefield, became the property of York County in the 1990s and was converted into an eighteenth- and nineteenth-century living history plantation known as Historic Brattonsville. Many of Brattonsville's original plantation buildings, including William Bratton's 1760s-era log house, were restored and opened to the public. In the early twenty-first century, the staff of the York County Culture & Heritage Museums (CHM), which operates Historic Brattonsville on behalf of the county, began a concerted effort to locate the battle site through historical and archaeological research. Staff historian Michael C. Scoggins began an in-depth research project that narrowed the probable battlefield site to an area that Bratton's descendants had described as being some 275 to 365 m (300 to 400 yd) due southeast of Colonel Bratton's colonial-era log house (Scoggins 2005). In 2006, CHM contracted the South Carolina Institute of Archaeology and Anthropology (SCIAA) at the University of South Carolina to conduct the first of several metal detector surveys of the site (Scoggins et al. 2011; Smith, Legg, and Wilson 2007).

The location of the Huck's Defeat battlefield was not rediscovered in a single field season. Numerous short-term field episodes slowly increased the number of battle-related artifacts, and, as important, identified areas where no battle-related artifacts occurred. The combination of these insights eventually revealed, and then confirmed, the location of the battle and Williamson's plantation house site.

The greatest difficulty in relocating the battle site was due to postbattle landscape change, the most significant change occurring because of nineteenth- and perhaps early twentieth-century cotton farming. That and subsequent timber harvesting appear to have combined to cause severe erosion of the battle site. Becky's Branch today is an intermittent drainage with steep vertical drop-offs, sometimes 3 m (9.8 ft) in height, from the naturally rolling landscape. The creek is fed by numerous erosion gullies formed and weathered by rainfall following the various land-clearing episodes. These gullies have been deeply cut into the natural slopes and terraces of the battlefield, and it is impossible to know which, if any, were extant at the time of the battle. Some of the gullies are also the result of numerous natural springs, and in fact, Williamson's house site proved to be located between two such springs, on a small knoll at the base of a ridge toe. Exacerbating the archaeological field efforts, the site is currently wooded in a combination of hardwoods on the slopes and slash pines along the hill crests, which possess a heavy understory of vines and bushes. Finally, there were several twentieth-century trash dumps consisting of metal items such as car parts and domestic trash. Systematic, complete metal detecting coverage of the area, key to identifying and interpreting the battlefield site, proved impossible until a concerted effort to remove small vegetation and metal waste was arranged by CHM.

The survey methods used were typical of those used to locate battlefields and included both metal detector reconnaissance and block searches. At the reconnaissance level, large areas were searched by several detector operators walking long transects spaced widely apart and designed to cover areas of up to several acres at a time. When battle-related artifacts were found, the area around the artifacts was blocked off using temporary flagging. These areas were then systematically covered by overlapping transects designed to examine and recover all artifacts from the entire block area. Each artifact was located using a GPS data logger. This method has proven effective and cost efficient at many southern battlefields (Legg et al. 2005; Smith, Legg, Wilson, and Leader 2007; Smith et al. 2009).

Whereas reconnaissance survey was difficult, the decision as to where to start the search was not. Using the testimony of veterans and their descendants, a light scatter of rifle balls was quickly discovered 420 m (1,377 ft) due east of the still standing Bratton house. After repeated transect passes, a total of 25 lead shot specimens were recovered in an area covering 3.3 ha (8.21 ac), with what appeared to be the core battle area covering an area of about 2 ha (5 ac). All but two of the lead shot were rifle balls or buckshot. One 0.69 in, 29 g musket ball was recovered, the size that would have been fired by a .75 caliber Land Pattern musket, the standard arm of British regulars. The final shot was a 0.58 in, 17 g ball from either a carbine or a pistol. Seventy-one additional metal items were recovered including a British halfpenny, a trigger guard fragment, and six eighteenth-century civilian buttons (Figure 3.3).

Clearly, the archaeologists had discovered a part of the battlefield, but was it the core of the battlefield? Repeated efforts to comprehensively or systematically cover 9 ha (23 ac) immediately surrounding the battle site, plus reconnaissance-level surveys on properties adjacent to and within an 804 m (.5 mi) radius of Historic Brattonsville, found no further evidence of battle-related metal or of any colonial occupation. The only evidence for any eighteenth-century occupation or military activity, besides the Bratton house, was the 3.3 ha (8.21 ac) described, which must be the general area of the Huck's Defeat battlefield.

Focusing in on this area for further investigation revealed archaeological features associated with Williamson's plantation home and the battlefield. This effort included not only systematic block metal detecting but also hand excavations and the use of ground penetrating radar, which was inconclusive. The results reinforced the conclusion that the concentration of lead shot and domestic materials was the site of Williamson's plantation home and the location of the Battle of Huck's Defeat.

Understanding the microlandscape of the site terrain greatly enhanced an understanding of how the battle unfolded. The 3.3 ha (8.21 ac) landscape consists of a ridge top and slope with two terraces dropping as much as 6 m (20 ft) northward into Becky's Branch, the lower terrace being a small 15 m (49 ft) square area bound on the east and west by two springs and their ero-

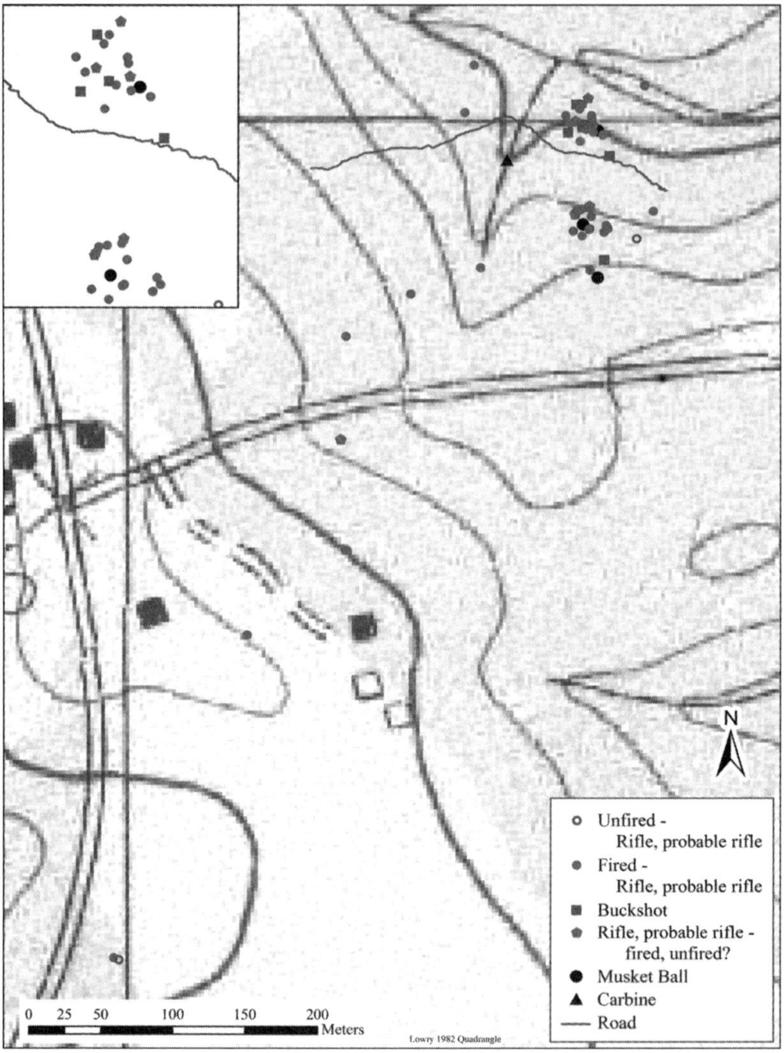

Figure 3.3. Distribution of lead shot at Williamson's Plantation battlefield. (Courtesy of the South Carolina Institute of Archaeology and Anthropology, Columbia)

sion gullies leading northward into the creek (see Figure 3.3). This lower knoll was where metal artifacts were found in a dense cluster and where eighteenth-century wrought nails were recovered that provide further evidence of the Williamson home site. The artifact scatter continued uphill to the upper terrace but ended on the level terrace face. Ten 1 × 2 m (3.28 × 6.56 ft) and three 2 × 2 m (6.56 × 6.56 ft) units were placed on the lower and upper terraces. Over 50 sherds of historic redware and some soft brick fragments were the only colonial-period ceramics recovered. No glass dating to the period was recovered. Thus the battle and domestic materials that are available to define the battle site consist overwhelmingly of metal items.

For all field efforts, the total number of battle-related artifacts included 45 specimens of lead shot, a trigger guard, and a sword pommel. The lead shot consisted of the aforementioned 29 g musket ball (British Land Pattern musket); two 22 g musket balls associated with the French Charleville musket and the standard arm of American regulars and Loyalist provincials; a carbine or pistol ball; and four buckshot. The remaining munitions were rifle or trade gun shot and buckshot. This array of ammunition has been found at other Revolutionary War sites associated with mixed engagements of militia, provincial, and regular troops (Smith, Legg, Wilson, and Leader 2007; Smith et al. 2008). Recovered camp and domestic metal items included forks, pewter spoons, flat buttons, shoe and knee buckles, horse shoes and horse equipage, wrought nails, and a door pintle, all consistent with a colonial-period time frame.

Although KOCOA landscape analysis was not necessary for discovering the battle site, its application allowed the researchers to reveal some defining features of the battlefield once its location was pinpointed. Instead of using KOCOA to find the battlefield, we used the battlefield site to locate landscape features that assisted in battlefield interpretation. In this case KOCOA was reverse engineered (see Sivilich, chapter 5, and Drexler, chapter 7). Some of the defining features for the battle included the Bratton house, the Williamson house, Williamson's Lane, Rocky Mount Road, and the rally point of the Whig force before the American attack. Working outward from the battle site, and with Bratton's house and the battlefield as the anchor, the other features were located. The Bratton house location was known, and the Williamson house site and the battlefield were found through metal detecting. Most significantly we discovered and mapped an intact remnant of Williamson's Lane, which consisted of a slight narrow trench-like depression. The road remnant trended east and west down the slope from Bratton's house, across an erosion gully, and passed just south of the Williamson house site. This trace closely fit the historical description of the road trace. Researchers had seen it, but until the battlefield was pinpointed, they could not be sure that the remnant was the historic road associated with the eighteenth-century battle and settlement.

Discussion

Huck's Defeat battlefield had been the site of significant relic collector activity before the field projects. The significance of this impact on the archaeological results for this battlefield is unknown. Usually archaeologists find one or two local relic collectors willing to share their finds once professionals turn their attention to the subject (Legg, et al. 2005). Through this interaction we can gain a rough idea of how much debris has been removed from the site. In this case, we found no collectors who would share their insights or collections. Certainly, some lead shot had been removed by collectors before our archaeological survey. In South Carolina, very few battlefields have not been visited by relic collectors. Nevertheless, we believe the sample recovered from the archaeological projects is representative of the battle and that valid observations and conclusions concerning that action can be drawn from the spatial distributions that they reveal.

The archaeological finds and their distribution strongly support the historical record. It indicates a short, sharp fight in which the American militia descended on the British camp and quickly achieved overwhelming suppressing fire from close range. Rifle shot from the Americans dominated the assemblage indicating little resistance by the British provincials and Loyalist militia. The densest concentration of rifle shot was found around the Williamson house site on the lowest terrace. On the second terrace above the house site was another concentration. A third cluster of shot consisted of a light spray (11 balls) beginning southwest of the site and ending west of the house concentration, but across an erosion gulley (see Figure 3.3). These distributions align with the historical record very well.

When the Whig militia split its forces into three groups and attacked the British camp from the front, left, and right flanks, the center group (lead shot indicating their line of attack) made first contact with the Loyalist militia camped on the hilltop and the second terrace (Figure 3.4). The Loyalists put up almost no fight, and resistance must have ended immediately. This is indicated by the lack of any rifle shot found north of the house or across the creek from the house site. Shot found in those locations would have been the result of overshots by the Whigs firing level at the Loyalists at the top of the hill. Had there been enough resistance to cause repeated fire by the Whigs, overshots would have occurred and have been discovered during the metal detecting survey. Instead, the concentration of lead shot at the Williamson house suggests that the Whigs quickly took command of the ridge and then stood along the ridge and fired downhill at their enemy along Williamson's Lane and around the house site on the lower terrace for the duration of the short battle. The lighter scatter of lead shot west of the house on the same topographic elevation as the Bratton house strongly suggests the British retreat route, which again could easily have been covered by the Whigs firing from along the ridgeline. This implies that the American force under Captain Read that was to approach from the west was as late to the engagement as was the eastern wing in the originally planned American

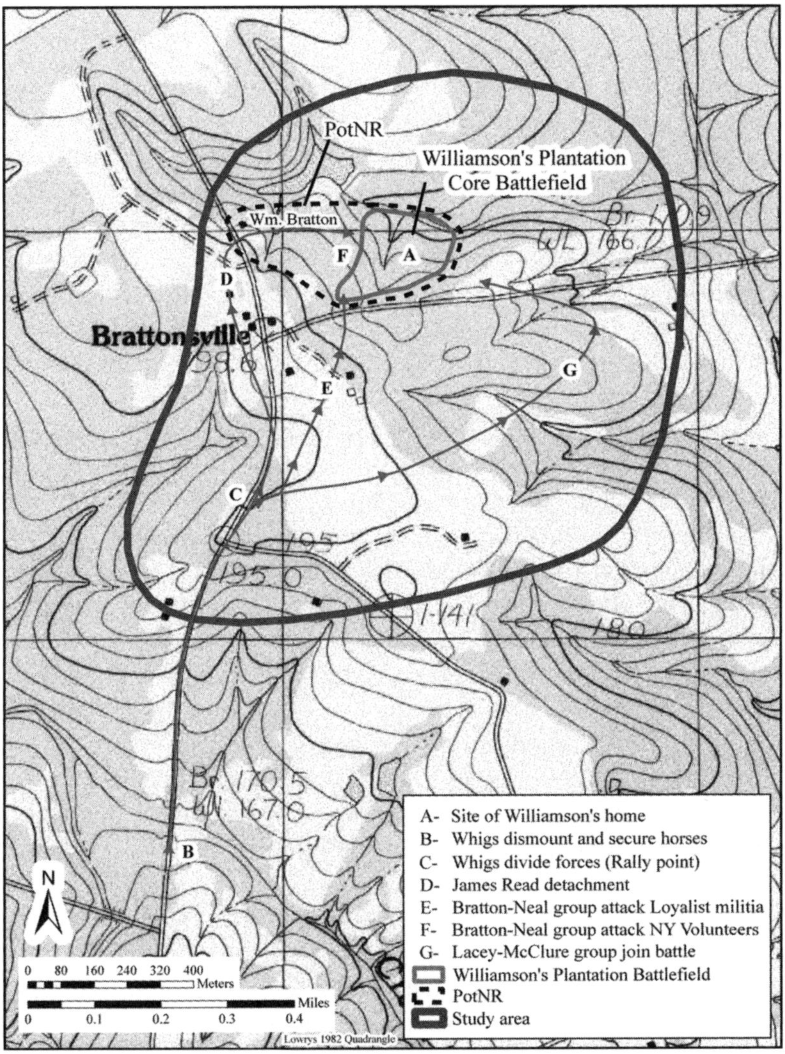

Figure 3.4. American attack routes at Williamson's Plantation. PotNR = potential National Register boundary. (Courtesy of the South Carolina Institute of Archaeology and Anthropology, Columbia)

three-pronged attack. Clearly, the American center commanded by Bratton and Neel quickly overwhelmed the British and Loyalists with minimal opposition before either of the flanking wings reached the battlefield. The archaeological evidence also seems to indicate that the Whigs did not pursue the routed British. Again, there is no evidence of firing to the north and northwest for up to 804 m (.5 mi) north of the battle site.

Conclusions

The Battle of Williamson's Plantation was an early, significant response by the Patriots after the fall of Charleston only two months previous. The victory aroused the backcountry resistance against the British invasion. For the British, it foreshadowed their eventual failure to subdue the backcountry and their ultimate defeat in South Carolina.

The Williamson's Plantation battle is a classic example of a partisan effort against a mixed force of regulars and militia. The Patriots hit the enemy with a quick dawn attack. It was planned as a three-pronged attack from left, right, and center, but in this case the flanking attacks were unnecessary, as the center detachment overwhelmed the enemy. Once the enemy was dispersed, the Whigs did not overextend themselves by following up the attack, where the possibility of increased casualties could occur. The raid is reminiscent of numerous partisan actions during the American Revolution such as Francis Marion's raid at Black Mingo (see Smith, chapter 4) or irregular warfare like the Native American raids against colonial setters (see McBride and McBride, chapter 1). Unlike Black Mingo, however, the Whigs at Williamson's Plantation felt confident enough in their victory to not immediately leave the battlefield and instead remained another day to bury the dead and transport the wounded.

The project also demonstrates the power of multidisciplinary analysis in historical archaeology, including using the memories of combatants to determine field strategies during the discovery phase of research. Likewise, this study provides an example of the value of repeated field efforts. Only after several surveys were the investigators able to define and refine the battlefield, including evidence of battle phases and single incidents within the larger engagement.

The Williamson's Plantation battlefield demonstrates that asymmetric warfare of raid, skirmish, and ambush, despite being a brief conflict with few participants, still leaves an archaeological signature, which can be read to reveal battle details otherwise lost to history. This signature, combined with the historical record, fills in important gaps in the record of episodic human behavior such as battle.

Acknowledgments

This research was funded by the York County Culture & Heritage Museums, the South Carolina Institute of Archaeology and Anthropology, and the National Park Service American Battlefield Protection Program.

4
Francis Marion's Partisan Community

STEVEN D. SMITH

During the American Revolution (1775–1783) in the region around Snow's Island, South Carolina, a close-knit colonial community strongly supported the American Whig resistance against the British. In August 1780, Lieutenant Colonel Francis Marion took command of this community of resistance consisting of both partisan warriors and civilians, and the region became Marion's legendary base of operations until late March 1781 (Bass 1959; Smith 2010). At that time the British destroyed Marion's camp, and the community regrouped and continued to support Marion and the American cause until the end of the war. This chapter examines Marion's partisans from the archaeological perspective of community theory, demonstrating that partisans engaged in asymmetric warfare requiring two essential elements for military and political survival. These elements are a secure base and a sympathetic population.

Francis Marion

Francis Marion (1732–1795), the "Swamp Fox," is regarded as the quintessential partisan of the American Revolution (Dederer 1983; Fitz-Simons 1995; Griffith 1978; Joes 1996, 2000; Keithly 2001:36; Laqueur 1976:19; Pancake 1985:110; Tierney 2006:42–43; Weller 1977). Military historians of guerrilla warfare compare Marion's skills to those of Mao Tse-tung (Griffith 1978:5–7). Even his enemies begrudgingly admitted his skills in asymmetric warfare. In fall 1780 Charles, Earl Cornwallis, commander of the British forces in South Carolina, wrote that "Col. Marion has so wrought on the minds of the People, partly by the terror of his threats & cruelty of his punishments, and partly by the Promise of Plunder, that there was scarce an Inhabitant between the Santee and Pee Dee that was not in Arms against us" (Cornwallis in Rankin 1973:115).

The following is typical of innumerable analyses of Marion and his place in the history of guerrilla warfare: "Francis Marion, in his late forties at the time, became the symbol of American guerrilla resistance. A teetotaling Huguenot, Francis Marion commanded a ragtag band of irregulars, including young boys and ex-slaves. They made their own clothes and weapons and shot their pistols with bullets made of pewter. Numbering no more than several hundred at its peak strength, Marion's force became skilled in night and dawn raids against British outposts. He and his men would ride as many as sixty miles [96 km] at a stretch along clay paths and marches of South Carolina in order to surprise a Tory or redcoat force. . . . In one particular escapade, . . . The guerrillas swooped down on the [British] column and freed the entire lot of Americans, an incident which made the 'Swamp Fox,' instantly famous" (Tierney 2006:42). Tierney recounts persistent themes whenever Marion is discussed in histories of partisan/guerrilla warfare and the Revolutionary War in South Carolina. These themes include: 1) proficient use of civilian or nonprofessional soldiers (partisans, militia); 2) small numbers of combatants; 3) classic use of irregular tactics such as surprise attack, ambush, and raid; 4) lack of sufficient ammunition; 5) outwitting of regular or professional soldiery; and 6) extreme mobility. Marion's tactics are still taught by the US Department of Defense (Wilson 2001).

There are two additional and closely related elements attributed to Marion's success that are crucial to any successful partisan/guerrilla or asymmetric campaign. First is the establishment and use of a secure base for supplies, training, recruitment, and rest. This support base must be "an area, strategically located, in which the guerrillas can carry out their duties of training, self-preservation and development" (Griffith 1978:96). "We [guerrillas] require a place that can also be used for conserving ammunition and food and for receiving and looking after wounded and sick soldiers. . . . As soon as we are the objects of the enemy's pursuit and attack, we withdraw there, and secretly hide, so as to await an opportunity to act or to begin resisting the enemy again" (Schram 1966:115). Isolated and/or rugged landscapes are the best location for such bases: "Geography is the guerrilla's ally, as much as it is the invader's enemy. Guerrillas know the land by heart; it is their home. They hide in the native mountains, swamps, jungles, farmlands, or urban sprawl, places where an outsider would seldom dare go" (Tierney 2006:17).

The second element is a supportive and sympathetic local civilian population, as Mao noted: "The more individuals there are among the people who support the guerrillas, so that the guerrilla unit can also maintain a communications network among the people, the easier it is for the guerrilla unit to find a hiding place. There are times when, in order to evade the enemy's pursuit and attack . . . a given guerrilla unit must be split up, each of its members being obliged to find a way to hide himself in one of the houses of the local population" (quoted in Schram 1966:116). Mao is supposed to have noted that the guerrilla must be like a fish in the sea, that is, be able to swim among the local population unnoticed (Turney-High 1971:263). During the American Revolution, the Whig households on and around Snow's Island were such a population: a partisan community—a community of resistance. As one historian noted, "Marion thus demonstrated an understanding that the true objective of guerrilla war is not the enemy army but the support of the population" (Waghelstein 1995:152).

Archaeological Community Theory

In this study, Marion's Snow's Island partisan community is viewed as a *community of resistance*, or a population of people acting asymmetrically, in concert, against a common enemy that is militarily and/or politically stronger (Smith 2010). More specifically, a *partisan community* is a community consisting of noncombatant civilians and full-time and part-time warriors, engaging directly (combat) and indirectly (material and political support) in guerrilla warfare against that commonly identified, stronger enemy. The partisan community led by Francis Marion, formed in the region around Snow's Island, South Carolina, during the American Revolution is offered as a case study for communities of resistance.

Community theory has had wide and diverse application in anthropological and archaeological research. The foundation of archaeological community research was the work of anthropologist George Murdock. Murdock and his colleagues defined community as "the maximal group of persons who normally reside together in face-to-face association" (Murdock et al. 1945: 29). Following Murdock, Arensberg and Kimball (1965) explained further that a community was defined by several elemental characteristics: shared territory (and environment), interpersonal relationships (kinship), and social control (government and religion). A more recent definition of community is "a co-residential, collection of individuals or households characterized by day-to-day interaction, shared experiences, and common culture" (Yaeger and Canuto 2000:2). For Kolb and Snead (1997:611), the community is where sociopolitical relationships are negotiated. These relationships are characterized by repeated interactions that help define who makes up the community. The community is, then, a social group where members recognize themselves to be members.

In a mobile society, for instance many of the Plains Indian buffalo hunters, the community was the band. Indeed, Marion's partisans, although part of a much larger society, nonetheless exhibited certain of the features of a band (Lawrence Babits, personal communication 2009). *Bands* are defined as "small, politically autonomous groups of twenty to fifty people, with an informal headman" (Keeley 1996:26). The band usually consists of a few extended families between which recognized, reciprocal obligations defined by kinship and marriage serve to mobilize community activity and identity. As will be seen, during fall 1780, Marion's partisan band consisted of only a handful of kin-related combatants operating without government support but secure within a larger network of kinsmen and in-laws.

The archaeological signature of community is "the physical locale [site] or cluster of locales where the members of a community lived, ensured their subsistence, and pursued their social functions in a delineable time period" (Chang 1968:3; see also Trigger 1968). At the time of the American Revolution, single sites (locales) were often domestic sites such as farmsteads, plantations, and mills that were occupied by individual households or families and dependent workers and slaves. Clusters of sites or evidence of "first-order, face-to-face" communities (Lipe 1970, 1992:3) could include a small village or town to which the outlying farms and work areas are tied and that can meet larger community marketing and service needs, security (militias), and government functions. Such communities were, in turn, tied to other more distant "first-order" communities that shared membership in even higher order social, economic, and political units at the level of counties, regions, and colonies. While acknowledging that there are concerns with this community-based perspective (Isbell 2000; Hegmon 2002), this study follows Chang and Trigger.

Archaeologists have used community concepts in the examination of resistance movements in the past. Brian W. Thomas's study of African American tradition and community in the South concluded that communities are formed by "individuals who recognize common interests, and are defined and contested through power relations" (Thomas 2001:18). Kenneth E. Sassaman, defining resistance as "action in opposition to structure, the opposite of compliance," compared Eastern European Gypsies, Old Order Amish, and Mexican Kickapoo and determined that there are "cross-cultural regularities found among traditions of resistance" (Sassaman 2001:219). Robert W. Preucel's study of architecture at Kotyiti, New Mexico, examined seventeenth-century Pueblo Indians in the Southwest (Preucel 2000). In attempting to resist the Spanish invasion, native leaders turned to a "pan-Pueblo ethnic consciousness" of shared traditions to unite a number of geographically defined communities.

Partisan communities have also received some attention from historians. Lindley Butler's (1994) historical study examined the Loyalist partisan community led by David Fanning in North Carolina. Fanning was the British equivalent of Francis Marion, and Butler described how Fanning's partisans relied on community networking much like Marion's did in South Carolina (Butler 1994).

Similarly, cultural anthropologists and social historians of Latin America have written extensively on local communities and modern guerrilla movements. These studies demonstrate commonalities between partisan communities of the past and guerrillas of the present. Dirk Kruijt (2008) studied three contemporary guerrilla organizations in Central America and noted that they all depended on the "local population as a vital source of material and moral support and as a reservoir of popular militias that served as allies of the guerrilla cause" (Kruijt 2008:93–94). Likewise, Timothy P. Wickham-Crowley found that peasant support was a "crucial determinant of the failure or success of rural guerrilla movements" (Wickham-Crowley 1992:51). As will be demonstrated, Marion's partisans operated in common fashion.

Methods

This examination of Francis Marion's Snow's Island partisan community required a multidisciplinary effort that integrated primary historical research, field archaeology, oral history, genealogy, deed and plat research, folklore, and a traditional review of secondary historical sources. The first step in this study was to establish that there was indeed a real community around Snow's Island before the war. The second step was to demonstrate that this community became a community of resistance during the war and to determine how it operated. Finally, the archaeological evidence of this community needed to be identified and interpreted.

Evidence of the human networking that defined the community before the American Revolution was established by demonstrating kinship and marital ties among its members along with the business, legal, religious, and governmental interactions that they shared. To accomplish this, genealogy studies, deeds, wills, marriage records, church membership rolls, lawsuits, land plats, and petitions were examined. Families that were related by marriage or kinship, intermarried, attended the same church, owned land adjacent or near friends and relatives, signed the same petition, or even sued each other to resolve disputes demonstrate a community in face-to-face interaction.

To demonstrate the existence of a partisan community of resistance during the war, audited accounts, militia rolls, primary accounts such as personal diaries and letters, and postwar pension applications were used along with written histories of Marion's asymmetric military campaign against the British. Archaeological evidence of the community was revealed by field survey and test excavation of known and survey-identified late eighteenth-century sites in, on, and around Snow's Island, South Carolina. As the presence of military arms and accoutrement provides the best evidence of partisan activity at a site, and as such arms and accoutrement are often metal, metal detecting by trained detector operators was the primary means of identifying sites associated with Marion's partisans. Although the presence of metal artifacts directed excavation, other artifacts of ceramic, glass, and other materials were often found in association and contributed to the evidence of encampment or site activity. Upon discovery of a suspected Marion site, more formal excavations were conducted to confirm the association and better understand its place in Marion's support system.

Supplementing these forms of research were local traditions and folklore of Marion-related events, oral history and personal family accounts, and the testimony of modern relic collectors concerning their knowledge of revolutionary era military and encampment sites and the types of artifacts found on them. Using all of these lines of evidence, a number of late eighteenth-century archaeological sites were identified that are arguably associated with Marion's partisan community. Jointly the information produced from these multidisciplinary sources clearly demonstrated that the success of Marion's partisans depended on their effective use of geography and terrain and their ability to mobilize the sympathy and economic and political support of the resident community.

Snow's Island Case Study

Snow's Island is physically defined as a landform of high land and swamp surrounded by three rivers or creeks: Lynches River (formerly Creek) flows into the Pee Dee River to form the northwestern boundary; the Pee Dee winds east and south to form the northern and eastern boundary. About a half mile upstream from the mouth of Lynches River, Clark's Creek branches off Lynches and winds southeast to the Pee Dee forming the western and southern boundary of the island (see Figure 4.1). The island is nearly 5 km (3 mi) from northwest to southeast, and 3 km (1.25 mi) across. Historical accounts indicate that Marion was supposed to have camped somewhere on this inland island.

Meanwhile, the Snow's Island region, inhabited by Whigs supporting the American cause and providing Marion's "community of resistance," is a much larger area. With the island in the center, the region that became a center of resistance to British occupation in fall 1780 is loosely defined as extending 24 to 32 km (14 to 19.5 mi) outward from Snow's Island (see Figure 4.1). The core, however, was composed of the families living on Snow's Is-

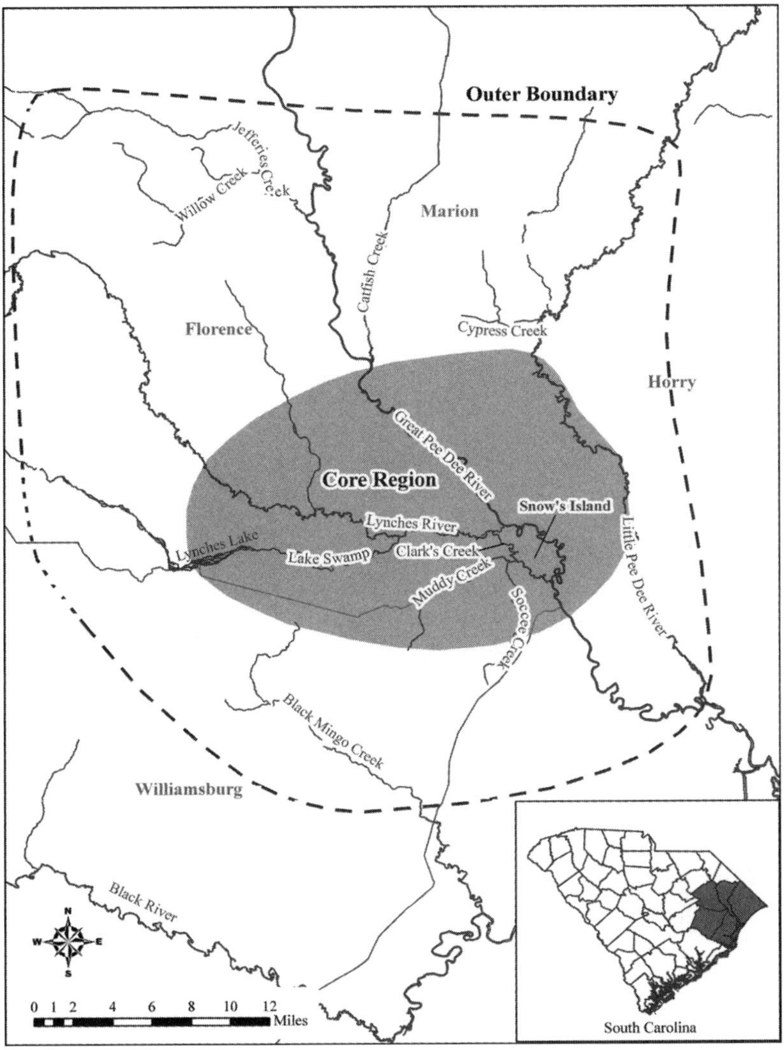

Figure 4.1. Snow's Island location, core, and outer partisan community region. (Courtesy of Steven D. Smith)

land and Britton's Neck across the Pee Dee River east of Snow's Island and those along Lynches River within that radius (Figure 4.2).

The Snow's Island region is part of the Pee Dee River drainage that originates far north in North Carolina and drains into the Atlantic at Georgetown, South Carolina. Around Snow's Island the land is generally low, swampy, and poorly drained. There are many oxbow lakes cut off by the ever-winding Pee Dee, along with numerous swales, small creeks, and ponds. Flooding and

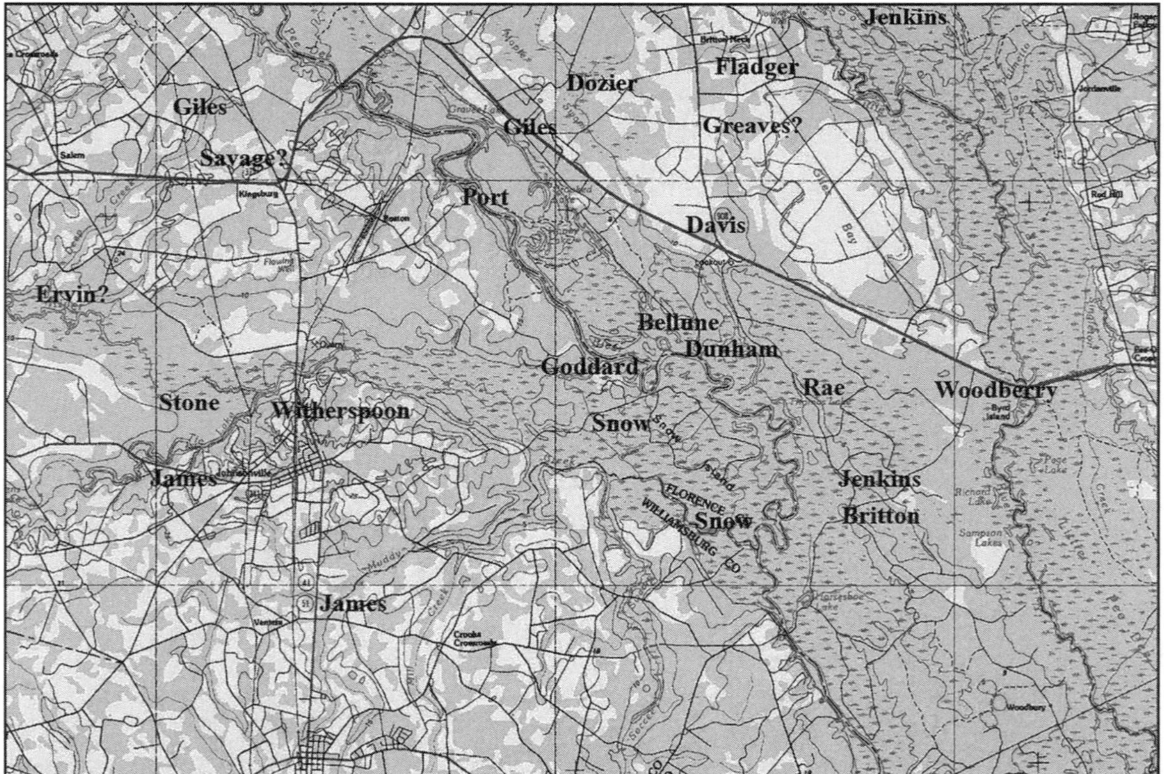

Figure 4.2. Location of families living in the Snow's Island core region. (Courtesy of Steven D. Smith)

relict marine formations have created dramatic and steep timber-covered sand ridges near the river bank. Modern dams and other major land modifications far upstream have had a great impact on changing the original riverine landscape and make reconstructing the colonial landscape extremely difficult.

This landscape was not especially attractive to the earliest colonial settlers due to its poor agricultural productivity. It was not until indigo dye became profitable in the South Carolina economy (second only to rice) that settlers from western Williamsburg Township, Charleston, and Georgetown arrived, beginning in the 1740s. By the 1760s settlement was widely dispersed across the region. These primarily English, Scots, and Scots-Irish settlers grew indigo and a little rice, and they raised hogs and cattle. The swamps were ideal for the foraging of hogs and cattle that could move freely and were rounded up seasonally for butchering. By the time of the American Revolution, the area was still sparsely settled (though in greater numbers) and remained largely forested. It was not necessary to clear large tracts of land as subsistence farming, stock raising, and harvesting indigo were all that was needed for a good life. Many settlers became wealthy plantation owners.

As settlement increased throughout the South Carolina backcountry, there was a growing need for governmental infrastructure. Initially all such power rested in Charleston, and that was much too far away to carry on business, adjudicate issues in civil and criminal courts, make land purchases, and process deeds, wills, and marriages. Furthermore, by the 1760s, the backcountry had become infested with outlaws and bandits, obliging the frustrated settlers to organize vigilante groups known as Regulators to provide security to the community. An anti-Regulator response finally forced the provincial government in Charleston to form new parishes and establish local courts and a legal presence (Linder and Johnson 2000:67). Though the Regulator movement was a failure, no doubt backcountry settlers learned a valuable lesson: rebellion could bring positive results.

The Snow's Island Community of Resistance

The Snow, Goddard, and Britton Families serve as typical examples of the many interrelated families that became part of Marion's partisan "community of resistance." Nathaniel Snow, a physician, came to South Carolina around 1696 and originally settled on a tract of land in Williamsburg Township, which included the land that was called Snow's Island by the time of the American Revolution. Many Snows later migrated west and settled on land south and west of the island. Snow's children included Mary, John, Nathaniel, George, Anne, William, and James. His daughter Anne married Francis Goddard Sr., whose son William owned a plantation on Snow's Island where Marion's main camp was supposed to have been located (Bass 1959:104). James Snow settled west of Snow's Island. He and his wife, Hannah, separately provided food and forage for Marion during the war. After the war, James, William Goddard, Austin Stone, and John James were named local commissioners. The latter two were prominent soldiers in Marion's Brigade.

There were two William Snows, one of whom settled on Snow's Island. The William Snow of Snow's Island was a prominent Anglican Church member and warden and was elected to the Provincial Congress in 1775. He sold some of his land to William Britton in 1775. The Brittons were prominent landowners across the Pee Dee from Snow's Island on what became known as Britton's Neck (the land between the Pee Dee and Little Pee Dee Rivers). William Snow witnessed William Britton's will. George Snow owned land on Muddy Creek that ran into Clark's Creek on the west boundary of Snow's Island. His land was adjacent to land owned by Moses Britton (Smith 2010:114–116).

The Goddard family genealogy is exceedingly complex with multiple individuals named Francis and William. However, the family is crucial to this study as one of its plantations on the north end of Snow's Island was where Marion is traditionally thought to have made his camp. Francis (I) Goddard had settled in Williamsburg Township before 1737. He had two sons, Francis (II) and William, and three daughters, Mary, Jane, and Frances. Mary Goddard married Timothy Britton, Jane married Phillip Britton, and their brother Francis married Anne Snow. The senior Francis (I) had originally purchased the land on the north end of Snow's Island but sold it to his son Francis (II), who eventually willed the plantation to his son, another William (II), who probably owned and farmed the land at the time that Marion arrived on the scene. William Goddard (I) married Elizabeth Britton and had one son, Francis (III). William (I) died in 1757. Elizabeth remarried Samuel Jenkins, who lived on Britton's Neck and would play a prominent role in confusing the British who raided the area during the revolution. Francis Goddard (III) also moved across to Britton's Neck on land inherited from his father but would die in battle in 1781 fighting under Francis Marion.

The Brittons of Britton's Neck, across the Pee Dee from Snow's Island, also had a complicated story, truncated here. They were a prominent family before the war and were actively involved in local politics, church life, and community service. The Britton brothers, Timothy, Francis, Phillip, Moses, and Daniel, settled the neck in the 1730s. As noted, several of the Brittons married Goddards and Snows.

Examination of as many as 46 colonial families at differing levels of detail resulted in a partial but representative list of pre–Revolutionary War families that lived in the Snow's Island region. Within this list, three tiers of community interaction and families were seen. The first level, or core families, consisted of 20 stalwart Whig families that lived in the geographical core of Snow's Island and/or the immediate surroundings. These were closely interrelated through kinship, marriage, land purchases, friendships, civil disputes, and church affiliation. They supported the revolutionary cause in general and Francis Marion's partisans specifically (see Figure 4.2). Often military-age members of these families joined the militia and fought directly under Marion in his brigade. The Snows, Goddards, and Brittons were core families.

The second tier consisted of 10 similarly related families. These were Whig supporters, many of whom fought under Marion, but they lived just beyond the geographical core. Some were related to core families. The third tier consisted of 16 families whose geographical location could not be pinpointed, except that they lived in the greater Snow's Island region (for instance, along lower Lynches River, although exactly where was undiscovered). Again, they were related by kinship and/or marriage to other community members and actively supported the revolution and Marion's command either by participating as partisan warriors or by providing Marion with hogs, cattle, rice, and other edibles.

During the early years of the war, 1775–1780, communities across South Carolina rallied to one side or the other. Areas around Orangeburg, Saxe-Gotha, and the upper Little Pee Dee in South Carolina saw the emergence of strong Loyalist communities, whereas areas like the lower Pee Dee, including Snow's Island, became Whig strongholds. No doubt, dissenters within these regions were harassed and sometimes forcibly removed. South Carolina communities became rarified, and as a Loyalist commander described the colony, it soon resembled a "piece of patchwork" (Gray 1910).

The Snow's Island community joined the resistance early. Community members first responded through participation in political resistance by joining the Continental Association, which declared opposition to British policy (Piecuch 2008:17). Names that would become familiar as Marion's Brigade officers or supporters included John James, Hugh Giles, John Witherspoon, Thomas Potts, Francis Britton, and William Snow. James and another Marion supporter, Thomas Port, joined the Provincial Congress. When news of the Battles of Lexington and Concord reached South Carolina, Port summoned the local male population to form militias, and he, Giles, and Potts were elected militia officers (Salley 1900:129). When the British fleet made its first appearance off Charleston Harbor in 1776, a call went out for assistance, and among many volunteers were Snow's Island community residents, including Francis Goddard and David Witherspoon (Smith 2010:133). Through the years between 1775 and 1780, the militias in the region were called out against the Loyalist groups. Yet as the war in the northern colonies continued with little action in the South, militias in South Carolina became bored with inactivity to the point that one local militia unit that had been sent to Cheraw, South Carolina, to guard an armory petitioned its conditions. The petition's signers included a long list of Snow's Island community residents, among them five Brittons, four Dunnams (Dunham), Francis Goddard, Thomas and Peter Port, and two Witherspoons. When the British returned to South Carolina in 1780 and attacked Charleston, Snow's Island residents were called to defend the city. Those responding included the Britton's Neck Regiment, commanded by Hugh Giles, who had become a colonel by that time (Smith 2010:138). Other Snow's Island families sent cattle, hogs, and fodder to feed the soldiers defending the city and their horses.

After the fall of Charleston, a British proclamation demanded all former rebels join in subduing those still in rebellion. In response, residents of the Snow's Island region began to think of more direct partisan resistance on their own. The strategic situation for them was grim. The British controlled all major market and political centers, towns, and villages to the south and west of Snow's Island including Charleston, Georgetown, and Camden. An American Continental army was forming in North Carolina, eventually to be commanded by General Horatio Gates, but for now they were on their own. They sent word for someone with authority to organize their resistance and command them.

Colonel Marion was sent to take charge of the Williamsburg militia, and on August 17, 1780, he arrived at Witherspoon's Ferry on Lynches River, just a few kilometers west of Snow's Island and in the heart of the Snow's Island community. A few days later, Colonel Giles and his regiment joined Marion's force. From this point until the end of March 1781, the Snow's Island region would be Marion's stronghold. His base of partisan/guerilla operations had the support of the Snow's Island community.

The British destruction of Marion's camp in March 1781 may have seemed like a defeat, but in fact it was a turning point. The American Continental army commanded by General Nathanael Greene entered South Carolina shortly after, and in cooperation with Greene, Marion's command and tactics evolved from that of asymmetric guerilla tactics toward a more conventional style of warfare. By September 1781, Marion's Snow's Island militia would stand in the front lines of a protracted formal engagement typical of classic eighteenth-century conventional warfare at the Battle of Eutaw Springs, the last major engagement in South Carolina.

Marion's Asymmetric Campaign and the Snow's Island Community

Marion's asymmetric warfare campaign against the British in fall 1780 began only a few days after he arrived at Witherspoon's Ferry. Under orders from General Gates to destroy all boats and ferries along the Santee River, Marion divided his partisans into two groups, sending one south toward the mouth of the Santee near Georgetown, and himself moving west and upstream with the other group. After raiding Murray's and Nelson's Ferries, he surprised a British convoy camped at Great Savannah Plantation. There he rescued Continental soldiers who had been captured at Gates's disastrous defeat at the Battle of Camden on August 16. He also learned that he was now the rebellion's sole, and unsupported, military presence in the northwestern part of the colony. He would remain so until General Greene arrived in South Carolina in December, but he would soon be left alone again, until April 1781.

That fall and winter of 1781 Marion became legendary as the "Swamp Fox" of the American Revolution, the actions of his militia keeping the British on the defensive in that region of the colony. His raids and ambushes included skirmishes at Murray's and Nelson's Ferries, Great Savannah Plantation, Blue Savannah, Black Mingo Creek, Tearcoat Swamp, Georgetown (twice), and the High Hills of the Santee, as well as a running defensive battle along the Santee River Road over the course of several days (Bass 1959; Smith 2010:148–186).

During this entire period, Snow's Island was Marion's base of operations: a place to hide, rest, recruit, train, and provision. The first definite evidence of this base camp in the historical record is from a reference dating to mid-November when bullets acquired from one Benjamin Port were to be taken to the "public camp at Johnston's [Snow's] Island" (Smith 2010:165). Even earlier camps had been established in the region, including those at Witherspoon's and Port's Ferries, and sometime later another camp was built on Dunham's Bluff across the Pee Dee River from Snow's Island. Besides "bullets," the Snow's Island community provided hogs, cattle, fodder, corn, rice, and slave laborers to the partisans around Snow's Island as documented in numerous audited accounts and receipts still extant in the South Carolina archives (Smith 2010).

Supplies were not only drawn from the community but also gathered in raids against the British and the Loyalist communities along the Little Pee Dee River. Marion controlled the gathering of these resources and their distribution. He not only planned raids with the goal of taking needed supplies, but he also saw to their redistribution to those Snow's Island residents who supported his partisan warriors. Salt, for example, critical to health, digestion, and food preservation, was one resource that Marion provided to his supporters.

The British were also active during this period. Regulars and provincials were sent into the region to root out Marion and his partisans as early as September 1780. One such raid, occurring while Marion was absent, cut a 24 km (14 mi) wide swath through the country, burning the homes of families in Snow's Island that were believed to support Marion's guerillas. This raid had the opposite effect of that intended and only served to inflame greater resistance and encourage partisan recruitment. When Marion returned to the region, he found support from the community was stronger than before. He immediately used that resentment to gather forces and attack a British outpost at Black Mingo Creek, about 16 km (9.5 mi) south of Snow's Island.

Although the specifics of every battle and skirmish fought by Marion's men during this period are beyond the scope of this study, the raids and campaigns followed a consistent pattern. Gathering strength and intelligence needed to plan an effective ambush while camped within the relative security of the Snow's Island region, Marion would make a rapid mounted excursion to surprise the British and then return immediately. Strategic circumstances often thwarted any concentrated efforts by the British to catch Marion's forays, but when they were able to locate a camp, Marion and his men simply dispersed into the swamps and reformed when the British left.

Marion's militia was left alone again in South Carolina in February 1781 when Greene's Continentals were forced to retreat into North Carolina. British forces remaining in South Carolina were finally able to focus on Marion and his partisans. At the end of March, the British launched a two-pronged attack against Marion's base at Snow's Island. One prong was launched from Fort Watson along the Santee River, the second out of Camden, South Carolina. Marion conducted a brilliant defensive campaign against the British advance out of Fort Watson, but the British troops advancing from Camden found the camp on Snow's Island and destroyed it.

About to retreat into North Carolina, Marion found that his fortunes had once again changed. Greene and his Continental army returned to South Carolina, and from that point on, the American forces began a long but sustained effort that eventually forced the British out of South Carolina in December 1782. After the fall of the Snow's Island camp, Marion no longer used it as a base of operations, but the community reformed and continued to supply Marion and Greene with cattle and hogs. Marion's period of asymmetric warfare was over; however, he continued to use guerilla tactics of raid, skirmish, and ambush as needed, including a brilliant ambush at Parker's Ferry in August 1781.

The Archaeology of the Snow's Island Community

The archaeological evidence of mobile partisans is generally ephemeral (see also Drexler, chapter 7). Further, diagnostic military artifacts that could be confidently attributed to guerilla activity would be rare as many of Marion's partisans were poorly equipped with civilian firearms and only a few diagnostic military-issue arms (see Scott, chapter 6). Nevertheless, by combining archaeology, historical documentation, oral histories, and conversations with local collectors, we can find ample evidence of the community

Figure 4.3. Eighteenth-century ferries surrounding Snow's Island. A = Witherspoon's; B = Port's; C = Potato Bed; D = Britton's. (Courtesy of Steven D. Smith)

and partisan warfare around Snow's Island in the form of landscape features and archaeological sites. Sites associated with Marion's partisan community of resistance include ferries, camps, fortifications, plantations, farms, and battlefields.

Access in and out of the Snow's Island region was channeled at the time of the American Revolution by four ferries: 1) Witherspoon's Ferry, west of Snow's Island, crossing Lynches River; 2) Port's Ferry, north of Snow's Island, crossing the Pee Dee River; 3) Potato Bed Ferry, east of Snow's Island on Britton's Neck, crossing the Little Pee Dee River; and 4) Britton's Ferry, south of Snow's Island, crossing the Pee Dee River (Figure 4.3). Whoever controlled these ferries also controlled access into and out of the Snow's Island core region. Today, the locations of these ferries are well known, either precisely (Port's and Britton's) or within a few hundred yards (Witherspoon's and Potato Bed). Witherspoon's Ferry site today is a bridge and boat landing. On the south side of the crossing, on nearby high ground and in a logical location for both camps and a domestic settlement, a late colonial site (38FL380) was found. The site was identified by a field survey that recorded the presence of historic ceramics. Unfortunately, the site was destroyed by the construction of a factory before it could be further evaluated for this study.

The west side of Port's Ferry is clearly evident as a road bed today. Metal detecting on the associated riverbank revealed three separate colonial period sites (38FL409, 38FL410, and 38FL411). Artifacts from the first locus included two side plates from British Brown Bess sea service muskets along with eighteenth-century buttons. At nearby 38FL410, a heavily looted site that collectors called the "muster field," archaeologists recovered five lead shot and two buckshot. At the third site, 38FL411, metal detecting revealed eighteenth-century buttons and a rifle ball.

Potato Bed Ferry is only known by its general location, having been destroyed by highway construction. Relic collectors have reported what they believe is an American camp including military buttons about 2 km (1.25 mi) from the logical location of the ferry site. Unfortunately, the site is now a clay mine. Britton's Ferry is marked by dirt roads leading to the crossing on both banks. At Witherspoon's, Port's, and Britton's, the ferries are located with one bank being a high bluff and the other side low and swampy with several possible locations for landing depending on the water level.

Since 1992, I have conducted short-term searches on Snow's Island proper in an ongoing attempt to locate evidence of Marion's legendary campsite, to no avail. Because the entire island is now forested and swampy, these searches could not be systematic and took place primarily on areas that logging companies cleared for timber. Whereas the location of Marion's Snow's Island campsite remains a mystery, a campsite has been located across from the island at a place locally called Dunham's Bluff (38MA207). At the time of the revolution, the site was owned by the Dunham (or Dunnam) family, which supported Marion's partisans. Pension applications indicate that the location was a rendezvous and muster ground early in the war and was used as a camp during the war. A partisan noted that his parents were taken to a camp "in the vicinity of the redoubt, which is situated on the right bank of the Pee Dee river opposite Snow's Island" (Munnerlyn 1833–1837). Indeed, only 150 m (492 ft) from this campsite, near the bank of the Pee Dee, are the extant remains of Marion's redoubt (38MA165). An early nineteenth-century South Carolina atlas by Robert Mills and other nineteenth-century documents depict Marion's camp to have been at Dunham's Bluff (Mills 1980 [1825]; Smith 2010:261–262). Thus, there are good arguments for and against this location being Marion's main "Snow's Island" camp, but either way, clearly it is at least one of Marion's dispersed camps that had been established around Snow's Island.

Metal detecting and exploratory trench excavations were carried out at the campsite and revealed a remarkable array of late eighteenth-century ceramic and glass artifacts, a 2 lb cannon ball, and, among numerous buttons, a block "I" silver sleeve button. This button is believed to be associated with the 1st South Carolina Regiment. Lead shot recovered was mostly rifle sized, and only one musket ball has been recovered (Smith 2010:270). Features found include trash pits and a possible palisade trench. The archaeological evidence indicates that this was both a domestic civilian site and a campsite.

The redoubt (38MA165) is the only undisputable military feature so far discovered around Snow's Island, except for a second possible redoubt at a place called the "Tanyard." The earthwork is triangular in shape, but only one corner of the triangle remains visible on the surface. Trenches excavated across the redoubt's palisade confirmed that it was man-made. A bullet mold for rifle shot was the only diagnostic artifact recovered at the site.

Marion biographer Robert Bass claimed, based on a few letters written by Marion in January 1781, that Marion's legendary Snow's Island camp was at William Goddard's plantation. Goddard owned land both on and off the island. Mills's atlas (Mills 1980 [1825]) depicts Goddard's plantation house across Lynches River from Snow's Island at the mouth of the river, which is consistent with historical descriptions and plat records. Metal detecting and block archaeological excavation at the site revealed two post-in-ground structures and associated late eighteenth- and early nineteenth-century artifacts (38FL282). The structure's post holes contained large sherds of creamware plates that had been heated to the point of the glaze melting. The current interpretation of the structures is that they are part of Goddard's post–Revolutionary War plantation, possibly rebuilt on the same location after being burned by the British raid in September 1780. The highly burned ceramics were used to fix the structure's posts in their post holes. No military artifacts were recovered at the site, although lead rifle shot was part of the collection.

One Marion-associated battlefield has been identified in the greater region. In September 1780, after the British had raided the Snow's Island community and burned partisan plantations, Marion returned from North Carolina and, in retaliation, attacked the British outpost at Dollard's Tavern on Black Mingo Creek, 16 km (9.5 mi) south of Snow's Island. This battlefield (38WG170) was located using metal detectors, and the collection included military artifacts including rifle shot, a gun trigger guard, a bayonet/sword scabbard clip, a gun side plate, and a ramrod pipe from a British Brown Bess musket. Brick ruins mark the tavern around which the British were camped. Across the creek on the north side, another camp (38WG171) was discovered by a relic collector who found rifle and musket shot, gun parts, and a large collection of buckshot. My archaeological metal detecting found more of the same.

Another Marion battlefield, at Blue Savannah, less than 16 km (9.5 mi) east of Snow's Island, was also metal detected, and a few specimens of rifle and buck shot were recovered along with eighteenth-century domestic material. Although intriguing, this is not enough to confirm the battlefield; however, historical document and plat research indicates that the archaeological team was in the right area and that the battle was a running fight over 5 km (3.1 mi). As such there may not be a concentrated core battlefield for this fight.

Archaeological investigations in the Snow's Island region have discovered a number of other late eighteenth-century sites that are intriguing and add circumstantial evidence for the activities of Marion's partisans. Two sites were discovered while searching for evidence of another redoubt on the east bank of the Pee Dee near Port's Ferry. The first, called Richardson's after the landowner, contained lead shot, buckshot, and eighteenth-century buttons, but no ceramics. It was located along a colonial period roadway that lay on the ridgeline running from Port's Ferry to Dunham's Bluff and is interpreted as a possible militia picquet post, which is a small, temporary guard post.

East of Port's Ferry, on high ground, metal detecting and survey collected two rifle balls, one buckshot, and late eighteenth- and early nineteenth-century ceramics. This site had been bulldozed before the archaeologists arrived, so that the associated landscape had been severely altered. The site is in the logical location for Marion's Port's Ferry redoubt.

Just north of the Dunham's Bluff campsite, and also located on the ridgeline and along the colonial road, a location was identified that gave up a random rifle ball and one buckshot. Interviews with local collectors indicate that this site in the past had been the source of a large collection of rifle balls and other colonial artifacts. The site is only 640 m (2,099 ft) from the Dunham's Bluff campsite and is interpreted as either another militia picquet post or an extension of the Dunham's Bluff site.

On a high hill, 5.5 km (3.25 mi) south of the Dunham's Bluff site, another large collection of colonial artifacts was discovered at a place known as Hickory Hill. This late eighteenth- and early nineteenth-century site is at the intersection of the previously referenced north–south ridgeline colonial road and the road leading from Britton's Ferry to Potato Bed Ferry. The site is probably one of the Brittons' plantation homes, possibly that of Daniel Britton. Its location would have been tactically important to any partisans attempting to guard the Snow's Island region, especially Britton's Ferry. Some 41 percent of the metal artifacts from this site were rifle balls and buckshot. As mentioned, an American camp near Potato Bed Ferry, reported by relic collectors, is only 3.5 km (2.25 mi) northeast of Hickory Hill.

Finally, there is the intriguing Tanyard site, located at the southern end of Britton's Neck. The site is located on a high hill overlooking the junction of the Pee Dee and the Little Pee Dee Rivers. There is nothing in the historical record to indicate that this site was tied to Marion's partisans, yet oral history, including that of local relic collectors, reveals that at one time there was a defensive work built by Marion at the site. Marion biographer Robert Bass, who was initially unconvinced that it had anything to do with Marion, ultimately changed his mind, writing a letter noting that there was still a small trench visible on the surface. The site was recorded in 1975, at which time the archaeologist recorded finding eighteenth-century ceramics and a "musket ball." Just before my survey, trees on the site had been harvested; however, Department of Natural Resources rangers stated that there were indeed trench works at the site. I visited the site and collected colonial period artifacts.

Overall, the archaeological signature of Marion's partisan community is diverse, varied, and broad. Although KOCOA analysis was not a focus of this research, like in Sivilich's study (chapter 5), KOCOA can be used here at a regional scale to inform our understanding of Marion's strategic use of the landscape. Each of the four ferries surrounding Snow's Island served as a choke point on a main Avenue of Approach onto the island. In this case, the ferries themselves served as Key Terrain to control or block access (Obstacles) into or out of Snow's Island or Britain's Neck, or to restrict the flow of traffic north–south down the Pee Dee River. Indeed, Colonel Robert Gray, a Loyalist officer, wrote a letter to his superiors noting that the construction of a redoubt at Port's Ferry with 20 men would control both Port's and Witherspoon's Ferries and communication south of the river (Gray 1780). Although the British ignored this advice, Marion recognized the strategic location of the ferry and did construct a redoubt at that location. Likewise, Marion constructed a redoubt at the confluence of the Pee Dee and Little Pee Dee Rivers to control water access to the region. At the battlefield level, the redoubt at Dunham's Bluff looks to be misplaced. It is located on the backside of an elevation rise that forms Dunham's Bluff. It obviously could not control river traffic. But its placement is perfect to control the road leading to the bluff and access to Snow's Island directly across the Pee Dee from the bluff.

Finally, Marion, like the Seminole (see Sivilich, chapter 5), scattered his camps, using not high ground but low swampland as Concealment for his vulnerable partisans. Although no example has been found on Snow's Island, the camp at Dunham's Bluff is an example of one of his detached camps.

Conclusions

The Snow's Island partisan community is a classic example of Mao Tse-tung's assertion that successful partisans and guerillas survive as a result of a sympathetic and supportive civilian population and a secure base of operations.

The Snow's Island community is also a classic example of how partisan communities and communities of resistance form and maintain their existence. The civilians around Snow's Island joined the Whig cause early in the war and provided food, forage, guns, blood, and slaves as laborers to the rebellion. When Marion arrived he provided security, salt, slaves, and leadership, linking the community's initial military exploits to the larger revolutionary political movement and strategic military effort. Together with direct and indirect community support, Marion waged an asymmetric war against the British until regular colonial forces arrived, and then Marion's partisans supported the regulars and eventually joined them in battles like Eutaw Springs.

During Marion's asymmetric campaign, Snow's Island became a generalized central place. In and around it, a number of small camps were established, some at strategically placed homes and ferry sites, allowing Marion to exploit a large resource base. The Snow's Island region was defined by the network and expanse of friendly supportive civilians and the contested border between them and neighboring hostile Loyalists. Snow's Island remained fairly fixed as a central place from September 1780 through March 1781.

Marion's Snow's Island base and its partisan community is an example of a larger resistance occurring across the southern states during the American Revolution. Within this larger community of resistance were enclaves of socially defined and physically bounded partisan communities like the Snow's Island region. These communities were the "piece[s] of patchwork" across South Carolina that a Loyalist officer described in 1781. They are also representative of Murdock's settlements of face-to-face interaction, cooperation or negotiation, and clustering; or of Lipe's "first-order, face-to-face" communities (Gray 1910:153; Lipe 1970, 1992:3).

Today remnants of this community remain on the landscape as either archaeological sites or landscape features that tie the past to the present. Archaeological evidence supports the case that Marion's partisans were widely dispersed across the landscape, and although they were guerillas, they did not shrink from constructing fixed defenses to strengthen their base.

Acknowledgments

This study was funded by the National Park Service American Battlefield Protection Program, the South Carolina Battleground Trust, and the South Carolina Institute of Archaeology and Anthropology.

5
KOCOA Considerations in Asymmetric Warfare

Education and Environment in the Second Seminole War, 1835–1842

MICHELLE SIVILICH

The Second Seminole War (1835–1842) was one of the first campaigns where graduates of the newly reorganized United States Military Academy at West Point, New York, were put into the field in positions of command. There they engaged with an enemy, the Seminole Tribe of Florida, who as inhabitants of Florida were desperately trying to preserve their lands and their way of life. The conflict took place between American soldiers, who were unprepared for the semitropical environment of Florida, and the Seminoles, who were highly prepared and had called Florida home for generations. These three Seminole Wars spanned more than 40 years, making this one of the longest conflicts the United States has ever engaged in despite being almost forgotten by history.

This chapter proposes a new perspective for understanding failed counterinsurgency efforts and uses the events of the Second Seminole War as a case study. This war provides an appropriate model because it was asymmetric in nature and also because its outcome was affected by many often-overlooked aspects such as the protocol of US officer training at the time, the nature of the environment in which it was fought, and the quality of topographical knowledge held by the opposing forces.

Consistent with many instances of asymmetric warfare, not only did both sides in the conflict have different skill sets when it came to conducting military operations in Florida, but the goals and means of achieving them for both sides were more complex than merely a struggle between Native removal and remaining. For the Seminoles the outcome of the war would affect their entire way of life. Their tactical decisions had to be based on preserving life. The judicial use of warriors in combat, the preservation of their limited population (including protection of their wives and children), the growing of crops needed to sustain their families, or the bargaining for needed supplies all played into their decisions in battle. In contrast, although settlers in Florida had to worry about the security of their families, soldiers, for the most part, did not. The army's tactical decisions revolved around providing security for dispersed farms and settlements, maintaining lines of commerce and supply, and finding the enemy in a hostile environment. Therefore, in considering the nuances of this conflict, this chapter proposes a novel way of utilizing KOCOA to better elucidate issues of institutionalized education, inadequate knowledge of environment, and the goals and motives of combatants in asymmetric conflict.

The Second Seminole War

The Second Seminole War was the middle of three wars that took place in Florida over relocation of the Seminole Indians. Involving some widely known military leaders of the time (Winfield Scott in 1836 and Zachary Taylor in 1838–1840), it was part of a long conflict that spanned over 40 years. However, these wars have been largely forgotten by history as they became overshadowed by other conflicts that took place during and after this time (Mahon 1985).

The Seminoles were a native group of several tribes that ranged throughout the lower southeastern United States. As pressures from Euro-American settlement increased in the more northern reaches of their territory, such as Georgia and the Carolinas, these various tribes retreated to their southern territories (Weisman 1989, 2007). The belief is that these tribes coalesced as a more cohesive group due to common threats from outside pressures and diminished numbers from introduced diseases (Craig and Peebles 1974). It is believed that the Spanish referred to them as Cimarrones (runaways or breakaways), a name that became Seminoles over time.

Initially, the Seminoles interacted and traded with the Spanish and British settlements (Fairbanks 1974), but this quiet time in Seminole history was disrupted in 1818 when General Andrew Jackson crossed the border between the United States and what was then Spanish Florida and launched several attacks, sparking the First Seminole War. This ultimately ended with the Adams-Onís Treaty of 1821, which ceded Florida to the United States. With this change in ownership came new pressures against the Seminoles. Although the Spanish had profitable trading networks with the Seminoles, the United States soon lost interest in continuing to trade with the native inhabitants. In fact, with the election of Jackson as president in 1828, his old rivalry with the Seminoles in the First Seminole War began to affect all tribal peoples in the United States, especially with the passage of the federal Indian Removal Act of 1830.

Tensions were high in the early years of American ownership of Florida. The government pressured the Seminoles to contain themselves within a reservation in the center of the state to prevent them from carrying on their coastal trading with Spain and Cuba. Then, with the Indian Removal Act, they were required to relocate west to a reservation in present-day Oklahoma (Weisman 1999). This relocation edict caused an outbreak of hostilities in which both settlers and Seminoles staged raids against one another. There were, however, also numerous other reasons for these tensions and hostilities, such as the question of whether the Seminoles were going to be allowed to take their slaves with them when they relocated (Mahon 1985).

Eventually tensions came to a head on December 28, 1835, when the Seminoles launched a two-pronged attack on US military forces in response to efforts to force them to relocate. The first, and most well-known, attack on this day occurred when a column of US troops of eight officers and 100 enlisted men, commanded by Major Francis L. Dade, was ambushed by a band of Seminoles. All but three soldiers were killed as they marched along the Fort King Road between what is today Tampa and Ocala, Florida (Mahon 1985, Sprague 2000 [1848]). The attack on Dade and his men was strategically timed to the Seminoles' greatest advantage; the place of attack was specifically selected so that the Seminoles could wait "in crescent-shaped formation with interlocking fields of fire," thereby maximizing the impact of their fire on the ambushed troops (Robbins 2006:38).

The second attack on December 28 took place when Indian agent Wiley Thompson and several other individuals were killed near Fort King (present-day Ocala). Thompson, who had been appointed to superintend the removal of the Seminoles in 1834, had ordered the imprisonment of Seminole leader Osceola at Fort King, even though the two were supposedly on friendly terms. Osceola, although not a hereditary chief, had earned his status as an important leader of the tribe. In response to this affront, Osceola personally led the attack that killed Thompson and six other people, including a sutler, two clerks, a lieutenant, and a boy (Sprague 2000 [1848]). According to Seminole leader Halpatter Tustenuggee (known as Alligator) the two attacks took as much as a year to plan. The goal was to maximize the impact on the citizens of Florida by instilling fear (Robbins 2006; Sprague 2000 [1848]). The attack on Dade's column and Osceola's killing of Superintendent Thompson have been acknowledged as the beginning of the campaign to relocate the Seminoles now known as the Second Seminole War.

The Second Seminole War went on for seven long years and consisted primarily of small skirmishes throughout the interior of Florida as the military tried to find and capture dispersed bands of Seminoles. Much of the military problem was caused by poor leadership and difficulties obtaining needed money and supplies from the US government. The turnover rate among commanding officers was high, with some tours barely lasting a year. Some reputations were tarnished as commanders failed to cope with and counter Seminole tactics in the field (Sprague 2000 [1848]). Eventually, as military pressure increased in central Florida, the Seminoles were forced to retreat south into the Everglades. The war was declared officially over in August 1842. Peace lasted for just over a decade before hostilities broke out again with the Third Seminole War starting in 1855 (Missal and Missal 2005).

Officer Education at West Point

By the turn of the nineteenth century, America had fought numerous wars and campaigns against Native Americans and as a result had an extensive history of both failure and success in irregular warfare, as demonstrated in this volume. Nevertheless, when it came time for the young nation to adopt a standing army, it appears that training at the new academy at West Point ignored this experience and instead turned to the rigid European style of military science focused on conventional warfare, such as that conducted in the Napoleonic Wars, ca. 1803–1815. This approach was largely due to Sylvanus Thayer, the first superintendent of West Point.

The United States Military Academy at West Point was established in 1802 as a loosely defined officer training system in which cadets would stay for any length of time and were of any age (McDonald 2004). In 1817, President James Monroe appointed Thayer to the superintendency, where he enacted policies of standardization, discipline, and rigorous testing based on what he had learned while in France at the École Polytechnique. It has been noted that "Thayer saw West Point as primarily a school of engineering and showed little interest in cultivating either liberal arts or strictly military courses" (Betros 2004:26). He modeled West Point after the French military institution, incorporating French military texts and treatises so heavily into the curriculum that the cadets were required to read French. In 1821 the first and second years of study were filled with math classes, including algebra, geometry, trigonometry, and calculus. By the third year, cadets were studying astrometry, physics, chemistry, and drawing. It was only in the final year that they studied fortification, artillery, tactics, and architecture (Board of Visitors 1821). Thayer's new curriculum shaped a four-year program that was designed to ensure conformity and continuity between graduating classes of US military officers. As one cadet wrote in 1822,

> If there is a means stronger than any other of cementing the union of the States, and perpetuating our government, it is the national Military Academy at West Point. To this institution, young gentlemen are sent

> from all sections of the Union. They come together with all the sectional prejudices, habits, and knowledge.... Their former habits, manners, and prejudices soon become extinct. They form a new character, a national character, which is no where else formed in the country.... They separate and are scattered to every part of the country; but their feelings are not separated, and their interests are not divided, and generally never will be. . . . From this source an uniformity of political principles and opinions and national and personal attachments will be formed and disseminated, which will bind together our States, and perpetuate our union, when without this cord they might separate. (Skelton 2004:29)

Perhaps Thayer's curriculum was a natural response for a new nation aspiring to European political recognition and acceptance. After all, the French had the leading military educational institutions in the world in the nineteenth century (Betros 2004). Their academies focused heavily on fortification and engineering and did not deal much with drill, tactics, or maneuver. This French connection ensured that the cadets approached problems using conventional formulaic methods: "Overall, the West Point approach to military science was narrow, technical and derivative.... Academically-trained military thinkers focused mainly on traditional fortification, tactics, and the particulars of military weaponry and equipment rather than on strategy or broad military policy. Moreover, they tended to accept uncritically European models rather than seeking flexible solutions to military problems within the context of American conditions" (Skelton 2004:33).

Although it may be impossible to show a direct cause and effect, the regimentation and standardization of the West Point education certainly influenced and shaped the tactical approaches used by US commanders in the field during the Seminole Wars. Of the 452 officers known to have served in at least one of the three wars, 15 served in the first, 285 in the second, and 146 in the third (some served in multiple wars). An analysis of forts constructed during the antebellum period through the Civil War shows that Florida, with 155, led the nation in the number of forts by state. Far in second place was Virginia with 70; all other states and territories paled in comparison. Of the Florida forts, 106 (68 percent) were built during the "Florida War" (Cullum 1868). A reasonable explanation for this is that during the Seminole Wars forting up was an automatic response and perceived solution by newly commissioned officers assigned to deal with the problems posed by the occupation of the Florida territory. For example, Lieutenant Henry Prince, who oversaw rebuilding Florida territorial Forts Forster and Dade, graduated from West Point in 1835 and one year later was serving in Florida.

If the lack of training in irregular warfare and the rigidity of the tactics of the day contributed to a lack of military success on the ground, perhaps part of the problem could also be attributed to a lack of command continuity. During the seven years of the Second Seminole War there were five top commanders. Most served barely a year. Scott served from February to December 1836, Major General Thomas Jesup from December 1836 to May 1838, Taylor from May 1838 to May 1840, General Walker Keith Armistead from May 1840 to May 1841, and Colonel William J. Worth from May 1841 to August 1942 (Division of Historical Resources 2015, Sprague 2000 [1848]). In some instances, failure led to court-marshal and tarnished reputation, as well as the publication of a flurry of angry letters either defaming or defending actions taken or not (Sprague 2000 [1848]).

The result of this lack of command continuity was that there was no time for a commanding officer to gain combat experience, learn from his predecessor's mistakes, and devise tactics adapted to the environment in which he was expected to operate. This problem extended down to the common soldier, whose tours of duty were short; militia and volunteer tours were even shorter. Although there was an American army during the entire war, it always consisted largely of inexperienced novices to the Seminole way of war and the humid, often lethal, tropical environment of Florida. The available archival record illustrates that similar tactics and methods were repeatedly employed by a succession of commanding officers, with similar levels of success or lack thereof (Sprague 2000 [1848]).

Terra Incognito

As is the case with many examples of asymmetric, guerrilla, or irregular warfare, one side of the conflict is a native inhabitant with intimate knowledge of the landscape and the opposing side is often nonnative and either unfamiliar with or less experienced operating in a hostile environment. In Florida this was certainly true. Most of the interior of Florida was unmapped and largely unknown territory. John T. Sprague, a participant of the later years of this war and publisher of a monograph on the history of this war, documents how many of the commanders conducted themselves throughout the war, noting their successes and sometimes their failures (although being active in the army, he was careful not to offend anyone). He describes some of the hardships commanders had to face in Florida: "Eustis and Lindsay, as shown by their testimony, had no competent guides, and could obtain none. Clinch's column was rather less badly provided in this respect, yet he had not a guide that knew any of the intricacies of The Cove, a Creatan labyrinth, held from the knowledge of the white man, as the sacred groves of the Druids were never entered except by the initiated. In short, all the difficult parts of Florida were, to the whole army, one terra incognita. Government gave me no topographical information, nor had any to give; and the booksellers' maps only afforded outlines filled up with unlucky guesses" (Sprague 2000 [1848]:143).

Florida was not only unmapped but unpleasant and unhealthy. The archival record is filled with soldiers' laments about the unusual, difficult, and often dangerous terrain of Florida—something they had not experienced the likes of before. Sprague often bemoans the difficulties that soldiers faced because of the Florida terrain, which was unlike other Indian removal campaigns:

> These results, trifling as they are compared with those of the Creek campaign, and with public expectation, are greatly beyond what we had any right to hope, when we consider the nature and extent of the country which has been the theatre of operations, and our utter ignorance of the greater part of it, even when we commenced the last campaign. . . . If our operations have fallen short of public expectation, it should be remembered that we were attempting that which no other armies of our country had ever before been required to do. I, and my predecessors in command, were not only required to fight, beat, and drive the enemy before us, but to go into an unexplored wilderness and catch them. Neither Wayne, Harrison, nor Jackson, was required to do this; and unless the objects to be accomplished be the same, there can be no just comparison as to the results. (Sprague 2000 [1848]:197)

> The progress of the Florida war from the 1st of January, 1836, was attended with large expenditures of money, and serious embarrassments. The climate, ignorance of the swamps and hammocks, and the treachery and activity of the enemy, baffled the skill of the most zealous and intelligent officers. (Sprague 2000 [1848]:96)

The movement of supplies and personnel proved nearly impossible. William Smith, another soldier who served in Florida, describes how difficult it was to move through the landscape:

> February 28.—Today we are out in the scrub and hammock land, up to our heads in briars and our knees in mud. I start on horseback, but finding it impossible (as the guide fortold) to progress with my horse, I have him returned to camp, and proceed a-foot. (Smith 1836:147–148)

> After toiling for hours to cut away the dense growth of trees, the very stumps left beneath the surface of the water, (and which therefore could

> not be trimmed lower) were so numerous that it became impossible to guide the horses through the tortuous clearing. They were accordingly taken out, and the men put in: or, to express myself in other words, the teams were unhitched, and the soldiers drew over the wagons by main strength, unassisted by horse power. (Smith 1836:154)

In this inhospitable environment, American forces still attempted complicated maneuvers to get at the Seminole. At one point, General Scott attempted to surround the Seminoles in the Cove of the Withlacoochee in north-central Florida. He had arranged for three columns to travel independently from different directions and converge on the cove on the same day. Without significant knowledge of the environment and terrain, however, the plan was doomed from the start (Mahon 1985). Each of the three columns was ordered to head to its assigned starting position. On a specified day, it was to begin its march into the cove. The center wing was supposed to flush the Seminoles out into the waiting forces of the left and right wings. With no accurate maps of the areas, however, the march proved to be longer and harder than imagined, especially for the left wing. Accounts by soldiers on the march declared that the plan was too complicated and the terrain too harsh (Mahon 1960:311). Without adequate supplies, means of communication, or knowledge of the environment, all three wings failed to meet their objectives and returned to their bases after having fought only minor skirmishes (Mahon 1985). As Robbins (2006:42) points out, "The conflict quickly fell into a pattern. Columns would march into the wilderness in search of the enemy, without a good idea where they were going. They thus presented an ideal target for the Seminoles, who usually had set up ambushes."

The semitropical climate also took its toll. Eventually, the United States abandoned the field during the summer sickly period, as described in this entry from the *Army and Navy Chronicle*:

> The Peninsula War.—The descriptions we have received of the fatigues of the march, the difficulties of the country, and the mode of fighting adopted by the Indians, give a discouraging account of the prospects of the final restoration of peace to the Peninsula of Florida. The scene of operations is a mere jungle of swamps and hammocks, which at this season begin to send out deadly, pestilential vapors, under the influence of which the strongest constitution cannot long stand. Disease has already disabled a large number of troops, and as the summer advances, a great mortality must infallibly take place. Bad roads, a country producing no supplies, toilsome marches, harassed perpetually by scattered parties of an enemy, whom they never see, but who

> watch them in the morasses, and, after firing a shot or two, disperse, and dive into the recesses, where they cannot be found, and no prospect of action to bring this dispiriting and exhausting struggle to a crisis, make this Florida war one of the most unpleasant that can well be conceived. (Army and Navy Chronicle, 19 May 1836, 3:318)

The Seminole Way of War

Whereas the US Army struggled with the environment, the Seminoles adapted their lives, including their way of warfare, to the Florida environment. Surprise, ambush, and raids were all part of the Seminole tactical doctrine (Butler 2001). Avoidance of any frontal assault, which risked unnecessary casualties, was paramount when attacking a stronger force. Leroy Eid's discussion of the Indian way of war, although referring to the 1700s, seems consistent with what was happening with the Seminoles in Florida in the first half of the nineteenth century (Eid 1988a). One tactic that was particularly successful for the Native Americans across the continent, and which they employed for a long time, was the half-moon battle formation used in the Dade ambush (Eid 1988a). This tactic probably began as a cooperative hunting strategy that also became invaluable in times of war and was used extensively by Native Americans (see McBride and McBride, chapter 1). As its name implies, this was a method of flanking the enemy without surrounding them completely, deliberately leaving an avenue of retreat or withdrawal. Added to this was constant motion. The Seminole attacks consisted of a running or "skulking" method of constant movement, which entailed more than simply staying hidden (Eid 1988a; Ferling 1980). Retreating when pressed was another tactic. A warrior would fire and retreat past his comrade who was loading and waiting for the enemy to follow (Eid 1988a:154). This method was called the Blackbird process and was used in the advance. The half-moon formation was traditional against equal foes, when the advantage was strong, but Native Americans also would close the flanks and surround the enemy. This contemporaneous account of the attack on Major Dade's command describes how the Seminoles purposely utilized the landscape to their advantage:

> Any one who has travelled, as we have done, the road from Tampa Bay to Camp King, will perhaps wonder why the Indians selected an open pine barren for their attack, in preference to the thick hammocks which skirt the Hillsborough and Ouithlacoochee Rivers. We think it can be easily explained. Powell, by his spies, had perhaps been informed of the movement of Dade's command, and sent a large force to cut them off, before they could form a junction with Clinch. Had

> the attack been made in the hammock, a number of the whites would have escaped under shelter of the woods. The object of the Seminoles was to destroy every man. Possessing, as they did, an overwhelming force, they would neither attack in these positions, nor attempt to surprise the camp at night, when the troops were well protected, and might have held out against a thousand Indians, or destroyed a large number of them. In the pine barren, the red men fought at equal and greater advantage, whilst all chances of retreat for the pale faces was cut off. (Smith 1836:76–77)

This half-moon formation, and the running method of attack described above, was neither random nor undirected but purposely utilized by experienced tacticians. Seminole chief Alligator was particularly known for his skill in battle, and "his knowledge of the country, and tactics in the field of battle, as displayed in advancing and retreating, in diverting the enemy and outflanking their extended lines, gave him an ascendency, and made him a dangerous foe" (Sprague 2000 [1848]:97).

Seminole battle tactics were both calculated and directed as was Seminole grand strategy. According to Coe (2016 [1898]:56–57), the Seminoles prepared well in advance of the first attack. He states that "the Seminoles now set about gathering their crops and making other preparations for war . . . they removed their women and children to islands in the extensive swamps situated in the southern part of their territory; their cattle were driven to remote ranges in the same direction; they laid in large stores of ammunition. Their supplies were considerably increased by the capture of a baggage train." Obviously, the Seminoles were approaching this conflict as a long-term insurgency, challenging the threat to their life and lands by a more powerful and well-supplied enemy. Removing their community from the enemy, they planned to apply pressure to their adversary through fast-paced engagements and more effective adaptation to the changing conflict situation. They were clearly successful: after over 40 years of conflict, while several thousand Seminoles were relocated and now make up the Seminole Nation of Oklahoma, several hundred remained in Florida (Mahon 1985).

KOCOA as a Matter of Perspective

Given the nature of the Florida environment, the necessity of the Seminoles' smaller forces to employ irregular tactics, and the Eurocentric tactics of the US Army, it would seem logical that each side had radically different perspectives on its approach to the use of the landscape for tactical purposes. Although neither side conscientiously applied KOCOA analysis in a system-

Table 5.1. Seminole War Examples of KOCOA Landscape Features

Feature	Definition	Military examples	Seminole examples
Key Terrain	Any landscape feature that confers an advantage	High ground, dry ground, roads, bridges, forts	Dense hammocks, hiding places, islands, inaccessible land suitable for farming, rivers
Observation and Fields of Fire	Any landscape feature that allows for surveillance of the enemy as well as the use of weapons	High ground, blockhouses, clearings	Locations within firing range of hitting the military but concealed behind trees or vegetation
Cover and Concealment	Any feature that allows for protection from incoming fire and obscures movement from the enemy	Forts, fortifications, breastworks, natural features such as ditches or ravines, etc.	Areas of dense vegetation, inaccessible hammocks, or islands
Obstacles	Any feature that impedes movement	Rivers, dense vegetation, swamps/wet areas	Fortifications and fixed defensive positions, military pickets
Avenues of Approach	Any route that allows for easy access to battle	Roads, trails, occasionally rivers when boats were available	Roads, trails, rivers

atic manner when confronted with a conflict situation, it is axiomatic that they did indeed apply the techniques of KOCOA from their own perspective. As Bruce Trigger (1989:261) notes, "human beings adapt not to their real environment but to their ideas about it, even if effective adaptation requires a reasonably close correspondence between reality and how it is perceived." Table 5.1 breaks down KOCOA according to landscape features.

In Florida, there were a variety of types of terrain that could be considered key. For the US military, it could be high, dry ground, necessary to observation and fields of fire, and lending an advantage in a conventional warfare setting. Likewise, when fighting irregulars, high ground would still be important for fixed defenses like forts. At the same time, Seminole Key Terrain could be an island or low hammock in the swamps that was inaccessible to the Americans. For the Seminoles, the goal was to stay hidden and in Florida, which was very different from the military's goal and thus led to different terrain requirements. Therefore, swamps and densely vegetated hammocks were for the Seminoles important Key Terrain that provided Cover and Concealment, but for the US military they were Obstacles.

Roads and river crossings were also Key Terrain landscape features in contrasting ways. Roads provided a lifeline of supply and rapid transit both for the civilian community and the military to and from logistical support bases and forts on the frontier. River crossings needed to be guarded against potential raids and ambush. For the Seminoles, such river crossings were also Key Terrain, in that they were known, fixed locations where the enemy would likely be vulnerable to ambush. Locations along roads, like narrows, were Key Terrain as ambush locations. Rivers provided Avenues of Approach for the Seminoles but were Obstacles for the US military unless boats were available. Sprague's writings (2000 [1848]) indicate that water as a barrier was a common theme; the Seminoles would disappear across a body of water or swamp, and the soldiers would be unable to continue their pursuit: "Driving the enemy before us for the distance of one and a half miles. . . . The regular troops and Floridians were pushed forward as rapidly as possible to the scene of the action, but meeting with an almost insurmountable obstacle in a difficult miry swamp, half a mile in extent, and from two to three feet deep in mud and water, were prevented from reaching it" (Sprague 2000 [1848]:164). In one case, soldiers who were wary of water (drownings were common) and refused to cross found out later that the water was shallow enough to have waded across: "The ford reached on the Withlacoochee by the troops, and which was so resolutely defended, was about three feet deep. The water was black and sluggish, deceiving those who ventured to approach. Had the contemplated charge been made, the river would have been crossed" (Sprague 2000 [1848]:166).

Additionally, water not only was a barrier to men and horses but made keeping their weapons and powder dry extremely difficult: "The nature of the country was such that soldiers could not even carry their cartridge-boxes. They were compelled to deposit them, with their muskets, in light boats, which they pushed before them through the mud for many miles to an island, where they found the Indians" (Sprague 2000 [1848]:196).

While the soldiers struggled, the Seminoles were very adept at using this environment to their advantage: "The enemy was vigilant and independent; he lived upon the wild productions, and knew the swamps and hammocks, while the soldier, as he waded them without food or rest, looked upon these fastnesses as his sepulcher, more than as a field of battle" (Sprague 2000 [1848]:166).

Cover and Concealment also had different meanings for the two opponents. The Seminoles used landscape features such as tall grass, trees, and ground cover for concealment and firing positions, whereas the military preferred manufactured forms of cover. For example, on several instances the soldiers in battle were ordered to create triangular log breastworks that they could lay behind for protection, rather than using the trees as the Seminoles were doing (Cohen 2016 [1836]; Sprague 2000 [1848]). This strategy was used by Major Dade when his troops were ambushed at the beginning of the Second Seminole War and again by General Gaines, who became trapped behind such a work for almost a week in early 1836 (Cohen 2016 [1836]). As the construction of such breastworks was used several times, it suggests that this was a standard tactic officers were trained to use in defense and that proper cover was considered to be a structure more substantial than mere trees.

When applying KOCOA archaeologists must also be aware that military leaders will likely weigh one aspect of KOCOA over another, based on the technology, environment, and culture of the period. For instance, they may decide that high ground, usually critical in eighteenth- and nineteenth-century warfare, though Key Terrain, is not as important as, say, controlling an Avenue of Approach. This can be seen in an examination of the location of four of the US military forts in the region. These forts, Fort Brooke, Fort Foster, Fort Dade, and Fort King, held a line that stretched between present-day Tampa and Ocala in central Florida along what was known as the Fort King Road, one of the most substantial roads through the interior of Florida at the time. Their locations, we might assume, would be placed to take advantage of Key Terrain features as they relate to the military tactics and texts of the time, in this case the highest local terrain. Yet in looking at each fort's location on a digital elevation model (DEM), we find that none of the four forts occupied the highest ground available (Sivilich 2012, 2014). Obviously, other factors were deemed more critical.

One explanation is that the constructed line of forts was less intended to fortify and provide security for an area and more to provide security for travelers and logistical support in the form of supply depots and way stations along an already established, economically vital road. In a letter to Major General Jesup, Colonel Foster states, "It is my duty to express to you my settled opinion, confirmed by time and experience that the system of Forts and Depots with light movable columns between them, composed of part horse and part Infantry is the only certain way of putting an end to this inglorious war without loss of many lives either of Horse or Man, and of causing much fatigue and suffering to the Troops" (Missal and Missal 2005:93).

The transportation of troops and supplies across the wilderness had been a constant problem for the military. Supplies were stretched thin across the landscape, and complaints about acquiring enough rations, animal fodder, tools, and other necessities were a constant theme in official dispatches. Maneuvering large baggage trains through the swamps on roads that were little more than cleared trails was a chronic problem, and yet the army could not function without the supplies they carried. Therefore, "forts," although having security roles, functioned also as storehouses and supply depots located at conveniently accessible locations along the main Avenue of Approach. This consideration was secondary to the need to have a superior defensive value during an outbreak of hostilities. Two of the forts were strategically placed at bridges to protect them. Certainly, Key Terrain meant more to the military than merely holding the high ground.

Also, these forts were typically very small (46–61 m [150–200 ft] on a side) and were not designed to house large numbers of soldiers. They were usually wooden structures that often consisted of a palisade, two blockhouses on diagonal corners, and possibly a few buildings inside. Most of the troops camped outside the fort walls, moving into the forts only when threatened.

That the forts served to provide supply and security functions for military and civilian travelers would suggest that for them to be most efficiently placed they should be easily reachable and that troops manning them should be able to effectively respond to threatened travelers. Yet these forts were more than 24 km (15 mi) apart, meaning that they were more than a day's march from one another. Messengers on horseback could travel more quickly between the forts, but troops on patrol or slow-moving wagon trains would often be obligated to encamp away from the security of the forts. During periods of hostility, troops bivouacked overnight between forts would have been vulnerable to attack, and the supplies carried in the wagon trains would have been particularly attractive to the materially stressed Seminoles. The Seminoles were quick to figure out the importance of the road system to routine troop movement and to the slow-moving baggage trains, and there were numerous instances of travelers on roads being attacked.

Florida was a war with no single frontline, and because the territory was still a largely unmapped wilderness it is likely that the high number of fortifications built during this time period was in response to the process of slowly conquering that wilderness. With such a large and difficult terrain to occupy, forces were spread thin, and resources were spread thinner, which is the likely explanation for why more forts were not built along the Fort King Road despite the need for them. Resources were needed more elsewhere.

Conclusions

Although neither the US military nor the Seminoles were specifically taught the KOCOA method, the logic of much of its application to the military use of terrain is well established or innate to the process of strategic and tactical decision making. The manner in which KOCOA issues is addressed by combatants in asymmetric military action can be expected to differ between sides, particularly in contexts that include cultural differences. Archaeologists must also be aware that military commanders necessarily chose among a multitude of factors when making decisions regarding which aspects of KOCOA are more critical. Key Terrain changes depending on mission. Controlling the high ground may be less important than protecting a road or bridge crossing.

Military tactics are defined by doctrine, which, in turn, is based on centuries of prior practices. Change in the early antebellum period happened very slowly, particularly at a time when the new US military was beginning to establish its identity. According to Steven Ross (1979), "parade-ground drill and battlefield tactics [in the United States] were virtually identical" to early European military practices. In fact, when the 1791 French training manual was updated in 1831, very little changed other than adding a section on how to fire from a two-rank line rather than from a three-rank line (Ross 1979). As the US military developed, it had just fought a war (1812) with the British, who were still seen as enemy. Accordingly, the official US Army training sought to counter European enemies, not Seminole. In the grand scheme of things, the Seminoles were considered to be merely inconvenient.

Eventually, in the mid-nineteenth century, American troops were trained to advance openly, rather than always in strict lines, and to use aimed fire (Ross 1979). Smith (1836:71) documents the use of this newer method when he writes, "We immediately extended as Light Infantry, covering ourselves by the trees, and opening a brisk fire from cannon and musketry. The former I don't think could have done much mischief, the Indians were so scattered." The military's lack of strategic and tactical adaptation to Florida's environment and to the nature of their enemy was one of its chief hindrances in carrying out the Second Seminole War.

In Florida, some small concessions were eventually made by the military that demonstrate that despite its strict engineering background, it could adapt to local conditions. For example, all brush and trees were removed from the area enclosing the forts, along the roadway, and key ambush points, allowing a clear field of fire against potential ambush. In addition, much of Florida Territory was abandoned during the sickly season of the hot summer months, a fact that may have given the Seminoles a respite for them to gather food and other resources. Despite these few concessions, the military was often inflexible when it came to strategy. Although this study only looked at fortifications, there are countless historical texts that discuss how each new commander in Florida would rely on tactics that had failed his predecessors. This provides an excellent illustration for understanding the powerful, yet overlooked, role that officer education plays in conflict. The asymmetric nature of this conflict where the native inhabitants have the clear advantage in navigating the terrain provides a stark contrast to outsiders' use of this same but unfamiliar terrain.

Additionally, there is a clear need for having separate terrain analysis models for each set of combatants that allows for a more nuanced and holistic view of these past conflicts. Soldiers and Seminoles operated in the same physical space but had radically different understandings and uses of that space. As conflict archaeologists seek to study asymmetric warfare, this becomes more important. Understanding the Seminole Wars as a complex set of interrelated features rather than a linear (or chronological) set of battles between opposing forces requires a more nuanced set of theoretical orientations than has been previously applied.

6
The Black Jack Battle of 1856 in Kansas

Asymmetric Warfare and Archaeological Investigations

Douglas D. Scott

Eastern Kansas was the scene of a series of armed clashes in the late 1850s that presaged the sectional violence of the American Civil War. On June 2, 1856, Free State militias under abolitionists John Brown and Samuel T. Shore fought with proslavery forces under Henry C. Pate at the Battle of Black Jack, the first pitched battle of the Bleeding Kansas era, which was characterized by bloody conflicts. The Battle of Black Jack is a seminal battle and, in many ways, is the harbinger of bloody and vicious fighting between Union troops and Confederate partisan forces that took place in the Trans-Mississippi West (Monaghan 1984).

The Battle of Black Jack is also an excellent example of nonstate actors with political and financial backing from a variety of disparate sources engaged in hit-and-run warfare tactics for spreading terror. The players had strong social and political convictions that deeply divided them and pitted them against one another. Both sides were well armed and, with the social, political, and financial backing from nonlocal sources, believed they were in the right. Each had set about to create a new social order to fit its view of how a society should be organized and maintained.

The issue of whether the Kansas Territory should become a free or slave state set the stage for the Battle of Black Jack. The participants in the action had popular support from both sides and, with their self-perceived legitimacy, had developed a tolerance or willingness to accept the dire consequences of disavowing the recognized rules of conventional conflict. This is one definition of an insurgency and asymmetric warfare. In this context, Black Jack is an excellent example of an instance of asymmetric warfare where the adversaries found it inconvenient to play by the rules and opted to play by rules of their own (Harvard University 2015).

A Brief History

The Battle of Black Jack took place on the prairie of eastern Kansas in part due to the presence of an important spring and the trace of the Santa Fe Trail that ran immediately to the north of the battle site (Figure 6.1) (Bondi 1904; Griffith 1925; Litteer 1998; 2002; Redpath 1860; Spring 1887; Townsley 1879). At this point, the trail forded Captain Creek immediately north of the spring, providing travelers a convenient place to stop, water their stock, and refresh themselves on their journey.

In May 1856, proslavery Missourian Captain Henry C. Pate, in command of Shannon's Sharpshooters, a self-organized militia, started for Osawatomie, Kansas. His goal was to capture John Brown, who was accused of perpetrating the notorious Pottawatomie massacre conducted on May 24 and 25 of that year. In their raid, Pate's men captured John Jr. and Jason Brown, although the elder Brown was in hiding. Soon after, Captain Thomas Wood arrived with a company of US dragoons and took charge of the prisoners.

Pate's company continued to the Santa Fe Trail, making camp on the head of a small branch of Captain Creek called Black Jack, also known as Hickory Point, five miles southeast of Palmyra, Kansas, on the edge of the prairie. That evening some of Pate's company rode to the town of Palmyra and captured several more prisoners. Six of his men attempted the same thing at Prairie City, but as it was Sunday, most of the targets were attending church. When a lookout gave the alarm at the militiamen's approach, the men rushed to their arms stacked in the corner of the church, ran outside, confronted the raiders, and captured two.

In the meantime, John Brown, with about 12 men, including three of his other sons, was following Pate's trail. Captain Samuel T. Shore, a Free-Stater who commanded the Prairie City company of Free State militia, assisted Brown in searching for Pate. They returned to Prairie City and encountered two men who brought news of Pate's camp at Black Jack. Learning the whereabouts of their enemy, Brown and Shore, with about 20 men, moved toward Black Jack. On arriving within a mile of the camp, they dismounted, left the horses in charge of two men, and dispatched two messengers for help—one to Palmyra and another to Captain Abbott's militia company some 12 km (8 mi) distant on the Wakarusa River. The remainder of the party divided, each captain commanding his own men, and marched toward the Pate encampment.

Pate had about 50 men under his command in the camp. When they camped, they had formed a hasty breastwork by placing four wagons in a line on the prairie edge probably east or northeast of Captain Creek and had pitched a tent behind, or likely east of, the wagons. Pate apparently de-

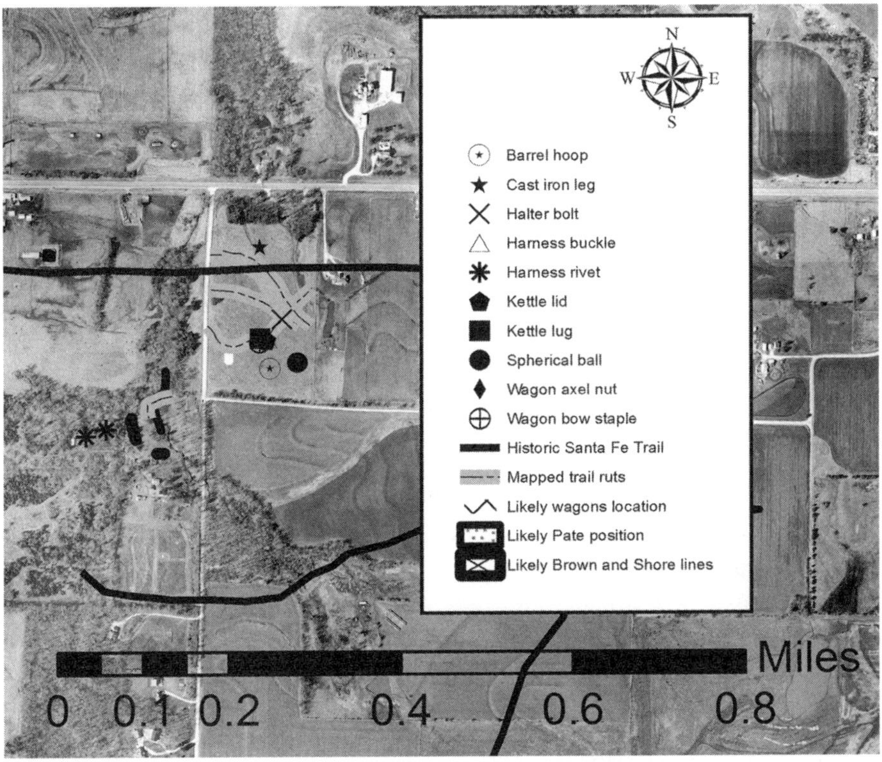

Figure 6.1. Map of Ivan Boyd Prairie archaeological finds showing Santa Fe Trail ruts. Note: the kettle lid and kettle lug are in the same location. (Courtesy of Douglas D. Scott)

ployed guards or pickets to the west, and one of these spotted Brown and Shore's men and gave the alarm by firing on the attackers. Pate rallied his men behind the temporary wagon breastwork in anticipation of an attack.

Realizing surprise was lost, Brown directed Shore to go the right and swing north to flank Pate's command. In the meantime, Brown proceeded into the upper part of a ravine or branch on the west side of Captain Creek. Apparently, this division of the forces was intended to catch Pate's forces in a crossfire, while using the ravine as expedient natural cover for the attacking Free-Staters. Shore's men approached the enemy over open prairie, probably from the south rather than the north, and unleashed a volley on the proslavery men at near maximum effective range for the weapons at hand. Brown succeeded in reaching the ravine, where he opened fire on Pate's wagon barricade. After a short time, Pate and his men abandoned the wagon breastwork and took shelter in the natural defenses of an entrenched eastern branch or ravine of Captain Creek. Shore's command, still on the open ground and exposed to Pate's fire, retreated across the sloping ground they had just crossed until out of range. Shore and a few of his men then joined Brown in the ravine.

The prisoners held by Pate were sequestered with a guard in the single camp tent, and when the firing began they lay flat on the ground to protect themselves as best they could. At some point in the fight, one of Pate's men went into the tent with the intention of shooting the prisoners. One of the prisoners, a Dr. Graham, at whom he aimed, succeeded in breaking away, receiving only a slight flesh wound or two. The others were apparently uninjured.

The fight lasted about three hours, during which time two Free-Staters and three proslavery men were wounded. Several of Pate's men began to disengage from the fighting and, one by one, gradually slipped down the ravine until out of range and left the field of battle. Pate's ammunition apparently ran low, and he finally sent one of his men and a prisoner to Brown's camp under a flag of truce, but Brown would not talk with anyone but the commander of the force. Pate finally came out, and after negotiations, Brown accepted an unconditional surrender. Pate surrendered only 21 men, the rest having deserted over the course of the battle. In addition to freeing the prisoners, the surrender yielded Brown valuable provisions, horses, mules, and other camp equipage, as well as a quantity of the plunder taken from Palmyra. Soon after the surrender, Captain Abbott and about 50 men from the Wakarusa arrived, augmenting the Free State forces. Later in the day still more men joined the force. The wounded were taken to Prairie City. Brown moved with his prisoners to Middle Ottawa Creek near Prairie City. He later met Colonel Edwin Sumner of the US dragoons and turned Pate and his prisoners over to him.

The Battle of Black Jack did little to settle the sectional differences between the Free-Staters and the proslavery faction. If anything, the battle inflamed the conflict and became a prelude to bloody and vicious fighting that continued for some time, giving the territory the undesirable sobriquet of Bleeding Kansas (1854–1861). These events were, in turn, a harbinger of the violence of the sectional fighting that occurred in the Trans-Mississippi West during the Civil War.

Site and Background

The battle site is in Douglas County, Kansas. Physiographically, Black Jack lies in the Osage Cuestas region of the tallgrass prairie (Mandel 2006:12–13). The area consists of long, low rolling hills separated by wide shallow valleys. The battlefield is located at the head of Captain Creek, which flows in a northerly direction eventually joining Wakarusa River, which flows into the Kansas River and eventually drains into the Missouri River. The rolling Ivan Boyd Prairie lies east of Captain Creek, the western side of the prairie draining into the lower reaches of the creek.

At the battle site, two short creek branches of Captain Creek intersect, one from the east and one from the west, to form the creek drainage. Several developed and undeveloped springs on the creek's west branch allow a luxuriant vegetation growth on the low-lying ground. From the creek, the land rises slightly to the east to become a rolling prairie. The creek banks are heavily vegetated with sycamore, elms, a few blackjack oaks, and a variety of underbrush that is often thick and impenetrable.

Before the archaeological investigations, the battle site was identified using a 1908 map drawn by William Connelley and 1925 sketch map drawn by George Roe. The earliest survey map of the area is the General Land Office plat, surveyed between November 30 and December 28, 1855, about seven months before the battle, by Thomas Connelley, who would later participate in the battle. Unfortunately, as this was a township and subdivision cadastral survey, little topographical information was noted. What was noted was sketchy and representative in nature only, but the plat did depict the location of the Santa Fe Trail running north of the site, as well as other trails, including one that appears to end abruptly just south of the Black Jack site.

Historical and Archaeological Investigations in the Black Jack Vicinity

In archaeology, one needs to know where artifacts are located and also where artifacts are not located. A primary research goal of the Black Jack battlefield archaeological inventory (Scott 2007) was to locate and define the limits of the battlefield, which has been recognized as a national historic landmark by the National Park Service. The first requirement then was to develop field procedures that allow coverage of the entire extent of the battlefield. The battlefield was relatively small and constrained and totaled some 8 ha (19.8 ac) of the Black Jack battlefield and about 14 ha (35.5 ac) of the Ivan Boyd Prairie. Most diagnostic, surviving artifacts of the battle and the Santa Fe Trail were either metallic or were associated with metal. Therefore, the archaeologists used metal detectors as an inventory tool based on the success of the technique at Little Bighorn Battlefield National Monument (Scott and Fox 1987; Scott et al. 1989). The use of metal detectors operated by knowledgeable people has overwhelmingly proven its value (Connor and Scott 1998; Espenshade et al. 2002) and is now a common practice employed in archaeological investigations of battlefields and campsites.

A controlled burn cleared the Ivan Boyd Prairie before the November

2006 inventory work. The metal detector operators used three meter-wide transects to cover the project area. The project discovered nine isolated artifacts and one artifact cluster (see Figure 6.1). Archaeologists recorded the specific location of the artifacts but also found clear evidence of the historic Santa Fe Trail. This consisted of a series of parallel depressions or ruts that run roughly southeast to northwest across the prairie reserve land. They are distinct on the ground and on aerial photographs as well. In addition, evidence of two trails or roads, more recent in time, was observed, leading to two stone commemorative trail markers. The recovery of nearly 100 aluminum beverage cans and lift-tab openers dating to the 1960s, concentrated at the highest elevation on the property, confirmed the recreational use of the reserve as a formal and informal picnic ground (Dolphin 1977:27–30).

A painting (ca. 1876) of the Black Jack battlefield site shows a wooden fence and gate in the foreground. Given the landform configuration seen in the painted image, it is likely the painting was made from a location on the Ivan Boyd Prairie. The wooden fence was located on the east side of the section line road that separates the prairie and the battlefield. Historic period artifacts recovered from the site were primarily cut nails and wagon parts. Some nails may relate to that fence. One other 8d cut nail, a 6d cut nail, and a hand-forged nail were found in proximity to one of the trail ruts but were widely scattered and could have originated from any number of sources. Given their association with the trail ruts, however, it seems prudent to suggest the nails may represent activity associated with travel on the trail.

Another artifact likely associated with Santa Fe traffic was a halter bolt found within the trail ruts. Well to the south of the trail ruts a barrel hoop fragment was found. A cast-iron kettle or grill leg located north of the ruts may have a trail association but could have other origins as well. The nails, halter bolt, and barrel hoop fragment are consistent with items known to be associated with nineteenth-century overland wagon construction (Gardner 2000). Archaeologists have recovered similar items at a site of abandoned wagons associated with the Donner-Reed party (Hawkins and Madsen 1990).

A single spherical lead ball was located on the southeast side of the Ivan Boyd Prairie. It is difficult to interpret the origin of this bullet as it could be from a weapon associated with trail traffic or the 1856 fight. Marks on the ball show low-velocity impact scars, suggesting that it was near its terminal velocity when it hit the ground. The firing direction cannot be determined; however, a .50 caliber ball like this would have a range of 366–550 m (1,200–1,804 ft), which would be consistent with it being an overshot from the Black Jack fight site, which is about 275 m (902 ft) to the southwest.

A cluster of artifacts was recovered from the higher terrain elevation near the southernmost stone commemorative marker. The cluster was sampled using the transect method employed throughout the inventory effort. The artifacts included cut nails, cast-iron kettle parts, and wagon parts.

A review of historical plat maps of the area found in the Kansas State Historical Society archives and dating from 1856, 1873, 1902, 1909, and 1921 (Beers 1873; Ogle 1902, 1921; Soudes 1909) do not illustrate any buildings on what is now the Ivan Boyd Prairie site at any time. Significantly, however, the 4d, 6d, and 8d nails found in the cluster are also known as common box nails, used to construct shipping boxes and packing crates of different sizes in the nineteenth century (Scott 1991:65–77). Rather than being from a frame structure, these nails likely represent remains of shipping boxes and packing crates, some possibly being wagon construction nails that were lost or discarded along the trail. The presence of the wagon bow staple, harness buckle, axle nut, and cast-iron kettle lug and lid attest to the trail activity association of this artifact cluster.

The unintentional and deliberate disposal and salvage of items during overland travel was a common occurrence during nineteenth-century wagon travel (Johnson 2006:6–10). Weight was always an issue, and it was common practice to divest a wagon of nonessentials and nonfunctional items as travel progressed westward to lighten the load. Discards were often left in piles along the overland trails so that others traveling by later could add to or make use of the discarded items (Hardesty 1997:77–79; Johnson 2006:6–10; Mattes 1987:51–52). As this artifact cluster was located on the high ground near a creek crossing of the Santa Fe Trail and a known watering stop, it probably represents an overland trail discard pile.

Other Information Sources and Collections

Nearby Baker University has a small collection of lead bullets purportedly from the Black Jack Battle Site in its archives (Figure 6.2). Archaeologists examined 41 bullets in this collection using forensic techniques of weighing, measuring, and examining them under a microscope. The clear majority of the bullets exhibited a white lead oxide on their surface that is consistent with patination of old bullets. Eight were of modern origin, but the others are consistent with pre–Civil War and Civil War manufacture. Three bullet types were in the collection: spherical balls, conical balls with hollow bases, and conical balls with solid bases. Additional manufacturing variations were present on the conical balls that allowed the bullets to be typed and dated to the mid-nineteenth century. The term *ball* used to describe the bullets is the standard military term for projectiles regardless of shape. The conical

Figure 6.2. Black Jack bullets from Baker University Archives. Note that most Minié balls shown have evidence of being fired and are deformed by impact at medium-to-high velocity. (a) .52-caliber Sharps, (b) .50-caliber country rifle, (c, d) .69-caliber Minié ball, (e) .69-caliber Minié ball with M1842 rifle ramrod mark on nose, (f) .577-caliber Enfield-type bullet, (g) .58-caliber Minié ball flattened by impact, (h) .58-caliber Minié ball exhibiting high-velocity nose impact deformation, (i, j, k, l) .58-caliber Minié balls exhibiting low-to-medium-velocity impact deformation. (Courtesy of Douglas D. Scott)

or spherical modifier refers to the projectile shape; conical forms are often attributed to varieties of Minié balls.

Several relic collectors came forward and shared their findings from their metal detecting of the Black Jack battlefield. One collector indicated that his group found 40 to 60 Minié balls of .58 caliber. Another collector told me that he had metal detected portions of the Ivan Boyd Prairie property and the Black Jack Battlefield Trust property about 1992. He indicated that he recovered about 12 bullets in total from the two properties. He recalled that the bullets from the Ivan Boyd Prairie were all spherical round balls. He noted that he found the bullets on the higher ground on the prairie property, and he thought that their points of discovery formed a rough line. He also stated that he had metal detected the area west of the east branch of Captain Creek and found three or four conical pistol bullets that he believed were Colt revolver types.

A metal detecting club from Kansas City detected the area on several occasions from the 1990s to 2006. The collectors recalled finding over 50 bullets including Minié balls in both .58 caliber and .69 caliber. Recovered spherical balls were smaller, around .45 caliber to .50 caliber. The participating members recalled that some bullets were located along the eastern branch of Captain Creek, but most of the spherical balls were located along the creek banks after the two branches joined at the north end of the current property boundary and on the adjacent property to the north.

The 2006 collection was available for examination and is a mix of recent and historic artifacts. Of particular interest are the hand-forged and cut nails, wagon parts, and lead bullets that the group recovered. Although their provenience is unknown, the historic wagon part finds are important as they parallel and are consistent with the archaeologically recorded wagon-related finds in type and numbers. With the exception of one Williams cleaner bullet and one .58 caliber bullet that was not introduced into service until late 1862 (Thomas 1997:211–240), all the lead bullets appear to be spherical lead balls. The calibers represented include: 1) two nominal .31 caliber balls, one with seven right land and groove impressions consistent with an 1849 Colt Pocket Model revolver; 2) one nominal .40 caliber ball; 3) two nominal .50 caliber balls, one with cloth patching impressions; 4) one nominal .69 caliber ball; 5) one O buckshot; and 6) five impact-deformed balls that are too damaged to determine caliber by measurement or weight. A single lead slug is of undetermined function but may be a .50 caliber (around 20 gauge) shotgun slug or a piece of lead meant for later casting as a bullet. These bullets and others from a private and an archival collection were analyzed using standard conflict archaeological techniques that are derived from forensic firearms examination methods (Scott et al. 1989; Scott and Haag 2009).

The 12 spherical bullets represent at least five separate calibers and gun types. One was from a Colt Pocket Model (percussion revolver), and a fired .69 caliber ball was from a smoothbore musket, possibly a variation of the US Model 1816 musket. The .40 caliber and .50 caliber balls probably represent common rifles or privately owned firearms. Any of these bullets could be associated with the Black Jack affair, but they could also be losses that occurred with travelers on the Santa Fe Trail. The fact that all appear to show some type of impact deformation, most moderate to slight, suggests that the bullets impacted near the end of their flight trajectory, perhaps approaching their terminal velocity. The range for these bullets would be around 450 m (1,476 ft) or 550 m (1,804 ft), except for the pistol bullets that would have reached terminal velocity after 275 m (902 ft) or 360 m (1,181 ft) (also consistent with their minimal deformation). If fired from the position held by John Brown's men and the Free State militia, on Captain Creek, they are within that range. Although this does not confirm their association with the battle, the possibility exists given the nature of the impact damage. Unfortunately, the lack of recovery provenience severely limits their interpretative potential. Nevertheless, the collectors' statements regarding the types of bullets found are consistent with most bullets noted in the Baker University Archives.

The Value of the Data Sets

The Baker collection, the bullet recovered during the archaeological work, and the bullets recovered by relic collectors and their recollections of where they were found, are consistent in types found and recovery locations, thus lending some credibility in using them as a proxy data set for providing some interpretations of the Battle of Black Jack. What is clear in the Baker collection, and confirmed by other recollections, is that most of the bullets found on the Black Jack Trust property are hollow-base Minié balls. This indicates that large-caliber, military style, rifled muskets were used in the Black Jack fight along with a few Sharps and country rifles.

Likewise, there is consistency in the type of bullets, spherical balls of moderate calibers, reported found to the north of the Black Jack Trust property. The spherical balls reported found and observed from the Ivan Boyd Prairie are consistent with the finds from the property just north of the trust boundary. These recovered and observed spherical balls suggest that a Colt revolver, a rifle or pistol of about .40 caliber, a rifle of .50 caliber, a shotgun, and a .69 caliber musket are responsible for the deposition of bullets on the Ivan Boyd Prairie.

To a certain degree, the Ivan Boyd bullets and those reported from the area north of the trust property are consistent with the types of arms attributed to the Brown command as indicated in the historical record. Those sources suggest that the combatants' arms consisted of a variety of shotguns, pistols, Sharps rifles, country rifles, and some old muskets (Herald Freedom, 15 December 1855, cited in Gridley 2001:24). The *Herald Freedom* on December 15, 1855, stated that Brown brought a quantity of arms with him, described in general as firearms, navy revolvers, and pole arms. The navy revolvers most likely refer to the .36 caliber Colt 1851 Navy revolver. The pole arms are likely pikes, a long pole with a blade and short crossbar on one end.

Arthur Birdwell's (1999) account of the fight mentions that the Free-Staters were armed with muzzle-loading rifles and shotguns as well as a few Sharps. August Bondi, a battle participant with Brown, recalled that when he joined Brown's company in late May 1856, Brown "carried a saber and a large-sized revolver; his sons and Thompson had a revolver, cutlass, and a squirrel rifle each; Townsley an old army musket; Weiner a double-barreled shotgun; Carpenter one revolver, and myself [Bondi] a flint-lock musket of the 1812 pattern" (Bondi 1904:281). Bondi's statement suggests that the Brown party armament consisted of personally acquired weapons and not government-issued firearms. If the spherical balls from the Ivan Boyd Prairie and the property north of the trust boundary represent shots fired by Brown's and possibly Shore's men, then there is good agreement between the archaeological and relic finds and the historical record. It is possible that the collector-recovered 0.63-inch-diameter ball represents either Bondi's or Townsley's old pattern musket, likely a .69 caliber.

In the only authoritative study of the weapons sent to Bleeding Kansas, Isley (1909) found in reviewing the New England Emigrant Aid Company records and other contemporaneous accounts of weapons provided to the Free State militias that before April 1856 about 200 Sharps rifles, five cannon, and 40 Millbury rifles and ammunition were delivered to the Free-Staters in Kansas. The Millbury rifles to which he refers are likely military-style muskets produced by one of the arms manufacturers in Millbury, Connecticut. Several firms in that area supplied military rifled muskets and accoutrement to the US Army under contract. They also manufactured firearms for various state governments and for the commercial market.

Watts (1998) has helped dispel the myth of the role played by firearms in the Bleeding Kansas era. He notes that John Brown in May 1855 called for the Free-Staters to be armed with a large Colt revolver and a Minié rifle with sword bayonet. A variety of pre–Civil War muskets and rifles, as well as imported military-style firearms, employed the saber-type bayonet (Gluckman 1965:179–182; Hardin 1987:86–147; Worman 2005:45).

The Baker University bullet collection is purported to be from the Black Jack property, although the provenience is not precise. The Baker collection, backed up by other collector recollections, clearly indicates that the firearms present during the battle included the Model 1842 rifled musket and possibly other older model .69 caliber muskets, Sharps firearms, country rifles in at least two calibers (ca. .40 and .45 caliber), and .58 caliber rifled muskets (probably Model 1841 rifles rebored to .58 caliber, although the Model

1855 rifled musket cannot be excluded). The important point of the collections is that no pistol or revolver-sized bullets are present nor are shotgun pellets or other large-caliber spherical balls.

If the Baker collection is representative of the bullets fired during the battle and deposited on the Black Jack property, then it may represent the firearms used by Pate's men during the battle as they fired at Brown's men hidden in the west branch of Captain Creek. Although not provenienced, if the conical bullets do predominately represent Pate, then his men were armed with a variety of surplus military arms and a few other weapons like the Sharps and country rifles.

The impact damage exhibited on most of the bullets is enlightening and informative. Most bullets exhibit some impact damage and are flattened on one side. This is indicative of bullets impacting at medium to high velocity either in yaw or being fired at a low incident angle. Such impact damage is consistent with the historical accounts of both sides using the ravines of the branches of Captain Creek as cover and firing at one another. Such a situation, rising up from the shelter of the ravine bank to fire at the opposing force, would allow for bullets traveling at medium to high velocity to impact at a low incident angle. Firing over a three-hour period would also foul the guns and cause fired bullets to become unstable in flight. Basal damage and side flattening in conical bullets are also consistent in this firing context. A few bullets exhibit slight nose damage, which is consistent with historical accounts of Pate's men firing from the open prairie to the west of the main battlefield. A few bullets exhibit significant nose damage from high-velocity impact. These bullets are consistent with the historical accounts as well, indicating shots fired at close range. In summary, bullet impact damage is consistent with the historical descriptions of the battle. However, without a clear recovery context, we must be satisfied with only the general correlations between the bullets in the Baker collection and the historical record.

One facet to determine is how many bullets may have been fired during the battle regardless of the weapon types present. According to the most reliable accounts, Pate had about 50 men at the beginning of the fight, Brown 10, and Shore 16. By the time of Pate's surrender, several hours after the fight began, Brown's command had dwindled to 15 men and Pate's to about 25 due to men being wounded or choosing to retire from the field. The primary account used to establish these numbers is a letter that John Brown wrote to the *New York Tribune*, which was printed on July 11, 1856 (Villard 1910:204–205). The total number of men involved is 76 at the beginning of the action. If each man fired three shots per minute for one hour (an impossibly high number given reports of Civil War engagements where men in sustained fighting rarely fired more than 10 rounds in a given battle), then the possible number of bullets fired might be as high as 13,680. Using the more realistic number of 10 rounds per man for the entire battle, then the number of bullets fired is 760. However, one must consider that several men were wounded, a number fled from the field, and some probably did not even fire their weapons (also a typical occurrence in combat even today). Griffith (1987:82) suggests that during the Civil War, a man in combat was more likely to fire his weapon once every five or 10 minutes during a battle. If this is true of Black Jack, and if we assume that the intense action lasted about three hours and that at least 50 percent of those there fired at this rate—essentially those left on the field at the time of surrender—then between 525 and 1,050 rounds could have been fired during the fight.

Assuming these numbers are reasonably valid, and considering the nearly 200 bullets known to have been collected from the site, then between 20 percent and 40 percent of the bullets potentially fired in the battle were removed from the site sometime in the past. This does not account for others who may have collected the battlefield who have never reported their finds.

Terrain Analysis

Given the lack of definitive evidence of the Black Jack fight in the area inventoried, it is appropriate to reassess the documentary and oral history resources from a different angle. The approach chosen was terrain or viewshed (areas of a landscape visible from a given point on that landscape) analysis that employs the power of GIS (Geographical Information System) computer-based programs. Cumulative viewshed analysis is simply a means to identify those parts of a landscape that are visible from a given set of points (Wheatley and Gillings 2002).

The first viewshed calculated represents the terrain view that Pate may have had if his wagons were on the prairie on the high ground just south of the Santa Fe Trail (Figure 6.3). This viewshed has a maximum range of 500 m (1,640 ft), which would be the maximum range of the majority of the firearms in the hands of the combatants. This postulated view shows that Pate had a fair view of the general surroundings but was somewhat limited to the south and west. If this view is reasonably accurate then Pate's rationale for placing pickets to the west is clear and demonstrates a good understanding of the tactical situation in which he found himself.

The second viewshed calculated postulates Brown's approach from the west, just south of the Santa Fe Trail as it runs toward Prairie City. Brown's terrain view as he moved from the crest of the ridge shows that he had an excellent view of the terrain for at least 500 m (1,640 ft), and he could easily see the wagons if they were parked on the prairie high ground. He also had

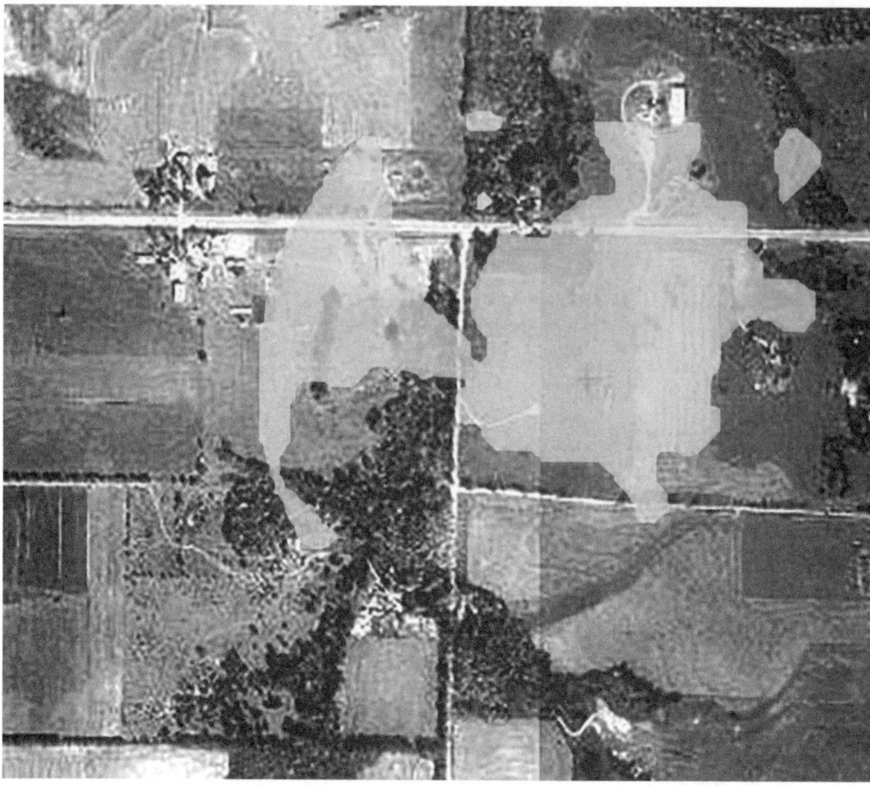

Figure 6.3. Terrain analysis viewshed or sight line from the postulated Pate wagon train site on Ivan Boyd Prairie. (Courtesy of Douglas D. Scott)

a good view of the creek and the ground from the north. If this location is correct, this also suggests that Brown had good tactical sense and was using the landform to his advantage.

Brown stated that as he encountered the Pate picket, he moved to the ravine, which would have been visible to him as he moved east and down the slope. Calculating the viewshed from the ravine using an observer height of 0.5 m (1.8 ft) above ground level still shows that Brown and his men had the prairie high ground across the creek in their sight and within range of their country rifles and old muskets (Figure 6.4). The ravine did provide maximum shelter for Brown, and even with fewer numbers he could provide suppressing fire on the wagons without exposing his men too much. This may well have forced Pate to abandon the wagons and move into the Captain Creek drainage.

Brown sent Shore to the right to flank Pate's wagon park. Assuming that

Figure 6.4. Brown's viewshed or sight line from the postulated western ravine of Captain Creek. (Courtesy of Douglas D. Scott)

Shore moved to the high ground and then down the slope toward Captain Creek and the prairie beyond, the calculated viewshed shows that Shore had a poor view of the prairie high ground and only limited views to the north and east. A second viewshed assumes Shore attempted to move around the east branch of Captain Creek and attack the wagons from the south. This viewshed analysis shows that Shore would have had good sight ranges in that maneuver. It is likely, however, that he could not have moved north without being observed, thus coming under Pate's guns and having less of a chance of achieving the goal of the flank maneuver. Either movement was possible. Shore may have come under fire from Pate's men at a longer range in this scenario, also consistent with the historical record, but he would still have had visual sighting of Brown as he moved into the creek's east branch. This would have given Shore the possibility of covering fire as he withdrew to Brown's position in the creek branch.

When Pate abandoned the wagon park for better cover in Captain Creek, he may have had as many as 40 or more men with him. The creek bed provided adequate cover for the men and a good field of fire. This shows that even tightly spaced individuals on Pate's side would have to spread out along the creek's east branch to the north. Brown and Shore, in contrast, could have easily occupied the west branch with the far fewer men in their commands.

Pate's men slipped away from the fight one by one or in small numbers, leaving him only 21 men to surrender. If there were men in Captain Creek and north of the east branch, then they could have easily slipped away north along the creek bottom out of view and range of Brown's men as reported. In fact, if the postulated deployment is correct for Pate's men then around 20 or more could have departed along the main creek bottom and maneuvered out of harm's way without Brown seeing them.

The viewshed and terrain analysis is speculative but does fit the historical description of the fight reasonably well. Combining the historical record, the archaeological and relic finds, and the terrain analysis shows that there is a reasonable fit between the three data sets, all independently derived. This correlation or concordance of data sets becomes a validation process of its own. It appears likely that Pate's command was encamped on the Ivan Boyd Prairie when Brown and Shore attacked from the west and south. Pate abandoned his wagons for shelter in the bed of Captain Creek and its east branch, while Brown and his men and Shore and a few of his men sheltered themselves in the creek's west branch.

Conclusions

The known relic collections attributed to the Battle of Black Jack at Baker University and from other collectors suggest that most of the evidence of the few hours of fighting by fewer than 80 men in 1856 has been removed from the main battle area. As frustrating as the loss of these items from in situ context may be to archaeologists, at least some artifacts were available to study and evaluate.

The Baker collection becomes an important data set as the only physical evidence that can be attributed reasonably to the battle site currently in a museum collection. A general expectation and perception gained from the historical documents is that the combatants were armed with personal weapons, shotguns, personal or country rifles, a variety of small arms, and a few of the latest Sharps breech-loading weapons. The Baker collection casts an entirely different light on the subject, assuming the collection is a valid representative sample of arms used at the battle, perhaps by Pate's command.

Using the Baker University collection data, a reasonable hypothesis can be drawn from the physical evidence to suggest that whereas Sharps rifles were present along with some country rifles, these were in the minority of firearms in combatant hands, at least those of Pate's command. Most firearms were military-style rifled muskets. If the bullet ratios noted in the Baker collection are representative of the number of firearms, then about 10 percent of the guns were country rifles or Sharps rifles, and 20 percent were older Model 1842 rifled muskets or other older model muskets subsequently rifled. These guns probably came from the state or territorial arsenal or had been privately purchased. About 70 percent of the guns were the latest model military rifled musket. These might have been privately purchased from various arms makers or obtained from stocks held in Missouri state arsenals. The US arsenal in Saint Louis likely issued these arms to the state militias under the various militia acts.

If the guns came from state arsenals, then this raises the question as to who authorized the issue. Each state or territorial governor had the authority to make the issues. Pate's command had some form of support either by private sources for the purchase of the latest model rifled muskets, or they were tacitly supported by Missouri state officials who had strong pro-Southern political leanings and had access to the state arsenals.

Pate's earlier attack on civilians and towns, and the taking of captives or hostages, very much fits the definition of terrorism—violence directed toward nonstate actors and noncombatants. Pate's armament suggests well-funded support or potentially tacit support from Missouri state officials who encouraged violence against Free-Staters in Kansas Territory, the goal being to drive them out of the territory or to terrify them into not voting pro–Free State.

Brown's command appears to conform to the traditional view that his men armed themselves with either guns or financial resources obtained from private sources. The motivation for Brown's attack on Pate's command clearly was the result of deep-seated political schisms regarding the place of slavery in Kansas Territory. In addition, Brown likely wanted to obtain revenge against Pate for his attacks as well as to secure the release of Free State–leaning prisoners they had taken, including his son.

The battle played out as a guerrilla-type scenario of unauthorized groups using hit-and run-tactics and employing ambush to wear down an adversary. The Battle of Black Jack's archaeological signature demonstrates how a single event can be reconstructed within the definition of asymmetric warfare, as fighters capitalizing on an imbalance in the knowledge of terrain and the use of new armament technologies to achieve their purpose. The archaeological record and the terrain analysis show how the fight played out. The physical evidence, the artifacts, demonstrate the type of weapons employed, and from that the larger picture of where and how those weapons came into the hands of the players is further explicated and integrates well with the historical record of the event.

The weaponry employed in the 1856 fight is diverse and may well be an archaeological signature of asymmetric warfare in general. Weapon diversity in the hands of nonstate actors has deep roots. This behavior can be seen in the American Revolution where both Loyalist and Whig irregular units used whatever firearms were available to them (Smith 2010, and chapter 4, this volume), and this is epitomized in the armaments of the minute companies that met the British at Lexington and Concord on April 19, 1775 (Fischer 1994). In more recent times, Mandzy (2016:199–202, and chapter 10, this volume) has demonstrated that Ukrainian partisans employed mixed firearms and ordnance against Soviet forces in 1945.

In the Civil War in the Trans-Mississippi West, especially in Missouri, irregulars, essentially guerrillas, were armed with a variety of guns, as is clearly seen in the archaeological record (Scott et al. 2014; Thiessen et al. 2008). Archaeological data recovery provides artifacts that can be assessed and interpreted in the context of an asymmetric warfare model. Archaeological investigations and interpretation demonstrate how simple artifacts have the potential to aid in reconstructing historical events and how they can aid in a more precise reconstruction of specific events, leading to a better understanding of the broader context of those events on the stage of history.

7
Jayhawkers, Bushwhackers, and Lay-Out Gangs

Archaeology and Asymmetric Warfare in the Trans-Mississippi Confederate Home Front during the American Civil War

Carl G. Drexler

The American Civil War in the Trans-Mississippi region is well known for the irregular warfare that scorched the valleys of the Ozark and Ouachita Mountains of Missouri and Arkansas. Wherever the US Army held sway, and where Confederate-allied local governments fell apart, violence erupted and endured, often even long after the cessation of formal hostilities in 1865. Memoirs by soldiers on both sides have been used extensively to document this conflict (e.g., Porter 1998; Williams 1982), although the irregularity, scale, and extent of the fighting mean that many events are lost to memory and many deaths went unrecorded. Secondary research in the Trans-Mississippi uses many of these sources to develop both popular and academic analyses of the war in the Arkansas and Missouri mountains (Bailey and Sutherland 2000; Barnes 2001). Fiction writers and film studios, too, have embraced the period and place, with, among others, Daniel Woodrell's (1998) *Woe to Live On* being adapted into the 1999 motion picture *Ride with the Devil.*

In 1993, historian Carl H. Moneyhon authored an article on irregular warfare in a different corner of the Trans-Mississippi, the Confederate home front in southwest Arkansas (Moneyhon 1993). Unlike the conflict in the relatively ungoverned mountain counties, southwest Arkansas was Confederate territory, served as the seat of Confederate authority after the US Army captured Little Rock in 1863, and never experienced lasting US Army presence or the concomitant disruption of local Confederate governments. Outside of the brief Camden Expedition of 1864, no blue-coated soldier set foot in the region unless he was a prisoner of war. This context offers a potential counterpoint to the disorder of the mountain counties and provides an opportunity to ask how Confederate authorities maintained law and order on the home front. In brief, Moneyhon's article painted a picture of a society gradually being torn along class lines, with violent opposition to Confederate rule becoming a major source of disruption to the war effort and to the peace and security for civilians (Moneyhon 1993, 2000).

Moneyhon's article focuses on the early 1863 period but makes clear that opposition in some form endured through the end of the war, and that it was both violent and organized. Although it was an influential article, there has been little follow-up research by historians. The paucity of documentation for Confederate Arkansas is likely a significant reason for this, as is the contradiction that the work poses to white Southern ideas about wartime service and patriotism. This chapter picks up on Moneyhon's work by discussing other indicators of antigovernment activity that are archaeologically identifiable, what modern conflict studies can do to illuminate this period of American history, and how this subject challenges our understandings of the Civil War on the western edges of the Confederate home front, particularly in the later stages of the war.

The Slow Death of the Confederacy in the Southwest

The home front of Confederate Arkansas in the latter half of the American Civil War is an important moment in the history of military doctrine because it is contemporary to the 1863 promulgation of the first systematic definition of different kinds of combatants. Section 4 of General Orders No. 100, better known as the "Lieber Code," classes other-than-conventional combatants as *partisans*, uniformed soldiers who operate detached from the main army; *armed prowlers*, civilians operating behind enemy lines to disrupt supplies, kill soldiers, or commit other depredations; or *war-rebels*, civilians behind enemy lines who rise against an occupying force (Lieber 1898). Outside of these categories lie criminals who exploit the disruption of social life to commit crimes. Of these, only partisans, when captured, had any hope of being treated as combatants guaranteed rights and prevented from being summarily executed.

The nearest term contained in the Lieber Code applicable to what we will see on the Confederate home front would be *war-rebels*, although the phrasing as used in the code suggests that such people are rebelling against an outside entity that has claimed and occupied their home territory. That label does not apply cleanly here, as the anti-Confederate agitation in question was against the home government. It is unclear to what extent, if any, the US government had a hand in fomenting opposition.

Much more applicable is the term *insurgent*, as used in modern military doctrine. In modern military jargon, an insurgency pits a conventional force against an opposing force that is not a member of regular armed forces, police, or internal security forces (Department of the Army 2008:2-2). The Army defines insurgencies as "the organized use of subversion and violence to seize, nullify, or challenge political control of a region. Insurgency can also refer to the group itself" (United States Department of Defense [USDOD] 2010, 2013:GL5). This definition is helpful for making a point about anti-Confederate activities in south Arkansas during the latter half of the war. Whereas anti-Confederate groups were armed and attempted to resist Confederate government attempts to control people and commandeer goods, they were not necessarily aimed at overthrowing the Confederate government or returning the area to Union control.

Using the modern meaning of insurgency in considering the events in the South in the 1860s gives us a better, clearer idea of what was going on. It also offers the opportunity to study how Confederate authorities combated insurgent activities. Modern scholarship gives us a language to characterize Confederate efforts, which then makes the context-specific aims of the insurgent more clear and cogent. It also expands and solidifies our knowledge of anti-Confederate agitation as initially presented by Moneyhon and encourages the consideration of multiple interpretations of Civil War–period sites throughout the southwest portion of the state, far removed from the actual enemy lines.

Irregular Warfare in Southwest Arkansas, 1863–1865

The Trans-Mississippi theater encompassed most of Louisiana, Texas, Oklahoma (known as Indian Territory at the time), Arkansas, and Missouri (Figure 7.1). It is an expansive area, but when compared to the eastern seaboard, it was lightly populated at the outset of the war. The area was long a home for the Caddo, Wichita, Osage, and a few other Native American groups, who were forced out in the early nineteenth century by the spread of American free white and enslaved black people following the Louisiana Purchase (La Vere 1998; Smith 1998). Through the first half of the nineteenth century, the border between the resident Native American and encroaching settler communities gave way to a cotton economy focused on plantation agriculture in lowland areas and smaller farms in less fertile uplands (McNeilly 2000). To this day, the area remains a thinly settled farming and logging economy, punctuated by a few cities.

During the Civil War, the Trans-Mississippi was not an area characterized by large battles between conventional forces in the way that Virginia, Tennessee, and Georgia were. More than 770 battles, actions, and engagements took place in Arkansas alone (Dyer 1908), but these were mostly small en-

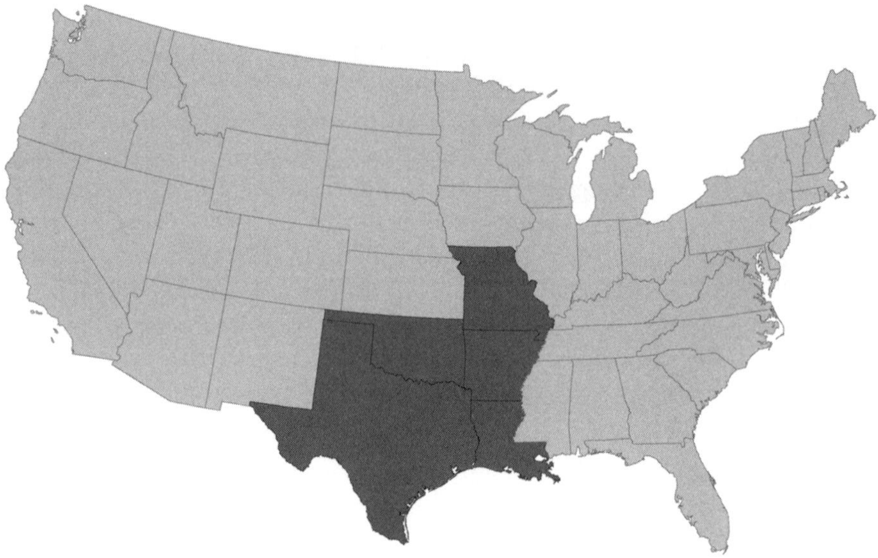

Figure 7.1. Trans-Mississippi South. (Courtesy of Carl G. Drexler)

counters between forces of limited size but high mobility. Indeed, the nature of the conflict in the Trans-Mississippi caused Collett (2007) to draw parallels to actions in Iraq and Afghanistan in the 2000s and 2010s. There were a few major battles, such as Wilson's Creek or Pea Ridge, but the preponderance of fighting in the area was on a much smaller scale. Within the region, Oklahoma experienced an extra level of strain, as intra- and intertribal schisms intersected with the larger conflict.

The Confederate home front in southwest Arkansas has not been heavily studied by historians. This is due in part to the fact that the vast majority of existing wartime documents from the state were generated by the US Army, which never had a significant or lasting footprint in this area. The Confederacy was not only less disciplined in recording local events but was also poorer in the means of making and storing such records. The works of Michael B. Dougan (1976), Moneyhon (1993), and others tell us that conditions were hard in this distant landscape, and that the people living there suffered mightily from the demands on and disruptions in their lives that were caused by the conflict. Dissension in the resident population became a problem by the end of the war for the Confederate administration, as it did elsewhere (Williams 2005).

The most comprehensive treatment of stress and war fatigue on the home front comes from Moneyhon's "Disloyalty and Class Consciousness in Southwestern Arkansas, 1862–1865" (Moneyhon 1993, 2000). The history of antigovernment violence in the area is sketchy, but it appears to boil over in early 1863 and was built around groups of former Confederate soldiers, disillusioned with the war effort, who came home to attend to their families and then bonded together for their mutual security and protection. These groups, often referred to as "lay-out gangs," ranged from 20 to 300 men and are representative of actions taking place in many Southern communities by the midpoint of the war (Williams 2005).

The Confederate authorities were determined to contain the growing violence and opposition and declared martial law on February 9, 1863. Partisan companies formed to chase down the lay-out gangs and "Jayhawkers" who were flitting around the mountainous counties near Arkadelphia. The use of the term *Jayhawkers* in many period documents is intriguing, as it is a term born in the Bleeding Kansas struggle of the 1850s to describe guerilla fighters who often clashed with proslavery groups, that is, "border ruffians," with whom proslavery Confederates were sympathetic (see Scott, chapter 6). As a number of Kansas units fought in Arkansas during the war, and some of the Federal soldiers may have taken part in that earlier conflict, the application of the term here may be a conceptual lumping together by locals of all those in armed opposition to Confederate authority into a recognized category of dangerous enemy irregular fighters. Much of what passed for Unionism at the time may have been more correctly understood as anti-Confederatism, as the members of many lay-out gangs were likely no great supporters of the North but were, rather, disaffected with the way in which they and their families had been treated by the Confederate government.

Numerous small clashes took place, with captured Unionist raiders often being hanged. The largest collision was at McGraw's Mill, near Arkadelphia, on February 15, 1863 (Sesser 2012). Unionist survivors of the fight fled to Fayetteville and joined the Union's 1st Arkansas Infantry. In the aftermath, under the aegis of martial law, dozens, if not hundreds, of anti-Confederate insurgents were hung by Confederate troops (Moneyhon 1993:238).

Though Moneyhon's narrative focuses on January and February 1863 and suggests that most of the insurgency was put down by March of that year, in fact, some level of opposition smoldered across southwest Arkansas for the balance of the war. The hillier counties north of the Red River were a focus of agitation. Civilian morale was low in early 1863, and it became significantly worse in the summer of that year, as the fall of Vicksburg, Mississippi, in July and Little Rock, Arkansas, in September not only cut the region off from the rest of the Confederacy but greatly reduced Confederate-controlled territory in the state. The resulting further dip in morale augured against Confederate efforts to maintain order on the home front. The diary of a John Brown (Swepston 2013), an elderly merchant in Camden, Arkansas, records numerous worries and incidents involving Jayhawkers from the summer of 1863 through the end of the war. In context, Brown appears to be using the term to reference disloyal raiders preying on their erstwhile neighbors, an example of the application of Bleeding Kansas terminology to wartime enemy irregulars.

Confederate diaries from southwest Arkansas indicate that anti-Confederate agitation was persistent and organized. William McPheeters (2002), a Confederate surgeon, noted in his diary while camped near Spring Hill, Arkansas, that two people, a woman and an elderly preacher, were brought into camp for their "Unionist" sentiments. In February 1864, Major General Sterling Price (1891) issued Special Orders No. 50, stating in the third section that "All disloyal persons who have been distributing incendiary papers or Federal proclamations, as well as deserters or persons at large without proper authority or who cannot give a satisfactory account of themselves, will be taken up and either turned over to the civil authorities or brought back with them." The forces engaged in taking up those individuals were instructed to act with "caution and audacity, as the enemy to be entrapped is both wily, unscrupulous, and well acquainted with the country."

These efforts do not appear to have resolidified order. On December 1, 1864, the *Washington Telegraph*, the only functioning newspaper in the area late in the war, issued a heavy-handed editorial in the hopes of establishing "friendly compromise, mutual forbearance, and mutual good [between] the Army and the People." It reminded its readership that Confederate failure would result in emancipation, miscegenation (which Southern whites would be helpless to avert), confiscation of Southern property to pay Northern war debts, and African Americans being made the equals of whites. All these points played on the racist fears of the Southern white population in an obvious attempt to beat back dissent against the Confederate government and illustrated the fact that some significant level of antigovernment activity was still a fact of life in the region. After all, what government seeks "compromise" and "forbearance" from a loyal, peaceful, supportive populace? This conveys a sense of the fractured nature of social life in the area at the time. This was not a unified populace, and Confederate concerns about organized insurgent opposition to the government were real.

Confederate Counterinsurgency in the Archaeological Record

The difficulty of researching this insurgency with the aid of archaeological methodology lies in the fact that there are virtually no known locations for the multitude of small run-ins between the Confederate Home Guard and regular units and the lay-out gangs and other dissenters. The skirmish at McGraw's Mill is perhaps the most celebrated of these encounters, but its precise location has not yet been identified and recorded as an archaeological site. Documents from the period hint at and allude to the act of armed opposition, but no known accounts exist from anyone engaged in the actual conflict. The nearest we have is the memoir of Dennis Haynes, a self-described Unionist who barely survived the war in Louisiana, having been shot in the chest and pursued across the state (Haynes 2006). We have no similar document from Arkansas. How then, do we chase this ethereal conflict through the dirt?

The use of irregular warfare and insurgency doctrine when combined with methods of archaeological research gives us a means to study this type of conflict, particularly if we learn from the research strategies developed by precontact specialists who have studied conflict among Native Americans. Such studies emphasize region-wide changes to settlement patterns and site architecture as indexes of conflict. They commonly note: 1) the construction of fortified villages; and 2) the establishment of depopulated buffer areas. They also commonly emphasize intervisibility between sites along with nonarchitectural data, including: 3) the rise of warrior iconography; 4) the presence of unburied bodies or bodies showing marks of perimortem trauma/mutilation; and 5) evidence for changes in trading patterns as hallmarks for endemic warfare that can be visible in the archaeological record (LeBlanc and Rice 2001). It is suggested that these indicators of conflict can be used within postcontact archaeology to develop region-wide models to study the impact of ephemeral conflict on communities and to start to identify and analyze conflict-related sites that can reveal evidence of insurgent conflict.

The luxury that precontact Southwest specialists enjoy is a mass of settlement patterning and artifact data, built by a century of widespread archaeological research the likes of which is absent for most postcontact conflict situations. The addition of the historical record in dealing with situations such as those in southwest Arkansas ameliorates some of these shortfalls formed by the absence of a developed archaeological record.

In Arkansas, some parallels to the precontact indicators of conflict noted are evidenced in historical accounts from the Civil War. The depopulation of much of the Ozarks and the mass migration of people to Texas between 1862 and 1865 (Baum 2008) essentially created a buffer zone between US and Confederate powers in the region. The US Army tried establishing "post colonies," communally defended enclaves of Unionists (Hilliard 2004; Hughes 2000), in northwest Arkansas late in the war. Hundreds of homes were burned, and isolated graves of people who died violent deaths speckle the landscape. In southwest Arkansas, the most direct evidence of conflict-related features on the landscape is a set of extant earthworks dug into the hills overlooking the Red River at Dooley's Ferry, a small hamlet at the time of the war.

Between 2007 and 2012, the Arkansas Archeological Survey, in collaboration with the College of William & Mary and the Arkansas Archeological Society, conducted field research at Dooley's Ferry, focusing on two elements of the wartime landscape: the community and the Confederate trenches that encircled it. Although the town no longer exists (Drexler 2013), the trenches remain and are easily visible on the landscape. In condition they range from shallow ruts snaking through the Dooley Hill Cemetery to substantial earthworks standing over two meters high. Subsurface testing focused on documenting the civilian community and the marks that the war left on it. This fieldwork indicated that half of the excavated structures appear to have been abandoned during or shortly after the war, suggesting that the disruptions brought by the conflict or its immediate aftermath caused a drastic constriction of the community (Drexler 2013).

In addition to the excavations on sites attributed to the civilian community living around the ferry crossing proper, the Confederate entrenchments that still ring the bluffs just east of the ferry crossing were mapped. These trenches wait archaeological testing, but the definition of their plan gives us an archaeological avenue to studying the factors that caused them to be constructed. The history of the use of trenches is complex and had some truly regional and national connections. The trenches at Dooley's Ferry were serious pieces of the military landscape and were also an interesting case of the intersection of race, labor, and internecine conflict during the war.

With respect to questions of the formidability and design of the Dooley's Ferry earthworks, Larry Babits has noted that Confederate entrenchments should be evaluated for their military effectiveness (Babits 2010), as trained military engineers were in short supply for the South during the war. The trenches at Dooley's Ferry were laid out by the same officer who designed the defenses of nearby Camden: Brigadier General Alexander Tappan. Tappan was a civil engineer before the war, and while rising to the command of an infantry brigade, he also appears to have mastered the arts of entrenchment and fortification. The trenches he laid out above the ferry are relatively small compared with other Civil War earthwork systems but are designed to make an efficient use of space while exposing few soft points in their defenses. They consist of three sections (A, B, and C in Figure 7.2) separated by roads and a patch of low ground that remains swampy and impassible to this day (section D is an earlier work sweeping the road through the bottomlands). The flanks of both the wing sections tie into what is, or was, low ground, meaning that they could not be flanked easily. Anyone trying to skirt the edge of the water bodies to get around them would face plunging fire from

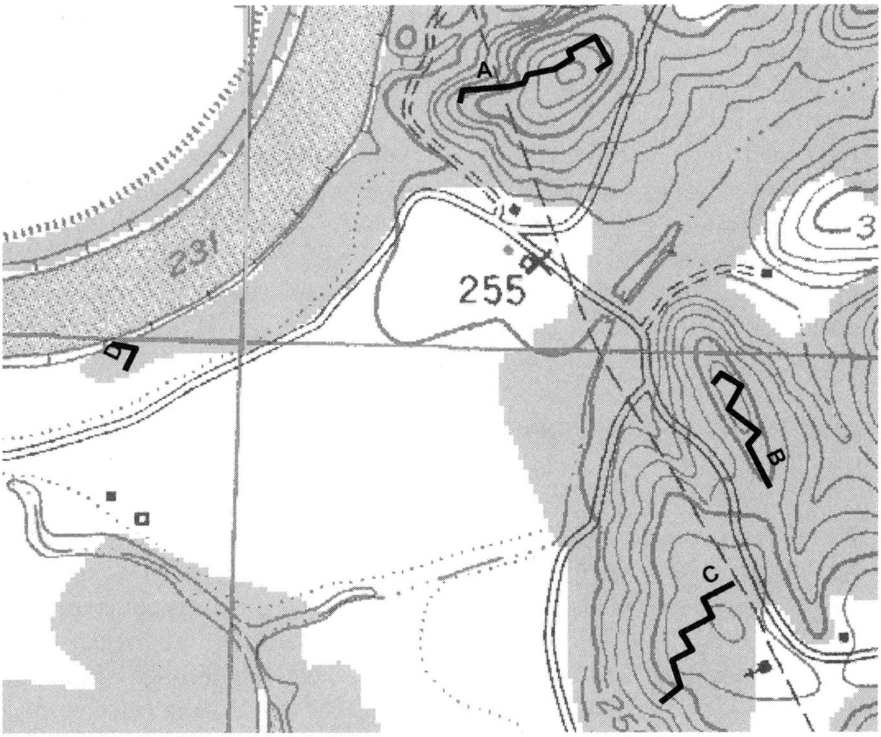

Figure 7.2. Trenches at Dooley's Ferry. Sections A, B, and C are shown. (Courtesy of Carl D. Drexler)

multiple sections of the earthworks that would have resulted in significant losses. The layout of the trenches shows a strong working knowledge of period defensive techniques as drawn from the writings of Sébastien le Prestre de Vauban (1737) and his various translators and interpreters, who significantly influenced military training and strategies of fortification design in the nineteenth century (Babits 2010; Halleck 1862; Mahan 1836).

Although the layout of the earthworks conforms to military doctrine, the timing of their construction offers an interesting story. Unlike the midwar constructed trenches at places like Tate's Bluff, to the north, the Dooley's Ferry trenches were built between late 1864 and early 1865, well after the threat of major combat operations in Arkansas. Following the early 1864 Camden Expedition, the US Army contented itself with remaining in Little Rock, displaying little to no interest in moving on southwest Arkansas. Major General John Bankhead Magruder, in command of Arkansas at the time, nevertheless ordered the construction of earthworks at Dooley's Ferry and Fulton, twelve miles upriver. These earthworks were far from the frontline.

Ostensibly, these could be used should a late-war push for east Texas be made, but this is not the whole story.

Earthworks, indeed, any kind of fortification, give an armed force control of terrain, both in the area immediately around it, and as a point from which the armed force can sortie to combat opponents. We know from documentary sources that insurgent lay-out gangs and bands of Jayhawkers were still active in the area at the time that the lines were constructed. This suggests the possibility that in addition to being defense in depth against a potential enemy US force, these trenches and the troops assigned to them were an attempt by the local Confederate government to control the movement of civilians and to combat insurgent activities in the area.

Archaeology and Confederate Counterinsurgency in Southwest Arkansas

Archaeology can build on Moneyhon's scholarship on wartime disruptions and disloyalty through the application of theory from precontact archaeology and modern military theory. Precontact archaeology's attention to site abandonment, the creation of buffer zones, militarization of popular culture, and changing trade patterns should be explorable through historical archaeological scholarship, and the depopulation of Dooley's Ferry points in that direction. That brings us closer to understanding the stress disorder of a population at war, but insurgency-specific archaeological resources are more ephemeral. Throughout the war, the only significant run-in that we know of between Confederate forces and insurgent Unionists was the skirmish at McGraw's Mill, although accounts indicate that many other small encounters were fought with lay-out gangs throughout the region. Some of these were instigated by Confederate Home Guard units seeking to break up the gangs, but the references to Jayhawker activities in Brown's journal suggest that they were far from reactive forces in the region. Whereas it is unlikely that they were organized into a statewide network, their size (bands of 20–300 individuals) suggests that they were not merely gangs of desperados. Finding these small encounters through archaeology is possible, though difficult.

Modern military scholarship offers another approach to the study of this period, which again ties into the landscape at Dooley's Ferry. It offers further support for seeing a substantial insurgency in the region late in the war by examining the history and archaeology of Confederate responses. Joint Publication 3-24, *Counterinsurgency*, includes a chapter on operational approaches to countering insurgent activity that emphasizes denying insurgents access to people and resources, almost all of which the Confederate government applied to Arkansans. Joint Publication 3-24 mentions population-control measures including "curfews, registration in a biometric database, establishing persistent surveillance systems, movement restrictions, travel permits, registration cards, and relocation of the population" (USDoD 2013:III–23).

Travel permits are an excellent example. Ex-governor of Missouri and Confederate major general Sterling Price's Special Orders No. 50, given in February 1864, subjected those persons "at large without proper authority or who cannot give satisfactory account of themselves" to arrest by military or civil authorities. That "proper authority" would be expected to come in the form of a pass or some other kind of authorization issued by a civil or military authority (Price 1891). The language used in Special Orders No. 50 closely resembles the strictures placed on enslaved African Americans during the antebellum period, forbidding their movement without a pass from their owner.

Travel restrictions are key to understanding the significance of the earthworks at Dooley's Ferry. The earthworks at Dooley's Ferry and Fulton were described as defense against a US foe that never gave any indication that it was planning on moving against those positions. They also sat athwart the two main points of communication and movement across the Red River between Arkansas and Texas. As such, they were the main avenues for civilian traffic back and forth, and investing them with troops gave the Confederacy control not only of unlikely Union action but of the passage of civilians and supplies entering the state or exiting it into Texas. Enforcing movement restrictions such as those reflected in Special Orders No. 50 would be made significantly easier by the fortifications.

We know that, at least late in the war, the earthworks at Dooley's Ferry and other places were far from being a simple bulwark against a Union threat. The detachment of troops stationed at the trenches was being used to keep civilians from leaving Arkansas. In a letter written from the location in May 1865, Leon Williams, a Confederate soldier in the 19th Arkansas Infantry, detailed to the commissary department, wrote to his sweetheart, Petronia McElroy, that the fortifications were being used to keep not only civilians but also soldiers from running to Mexico, a destination to which he felt drawn.

Period accounts document that thousands of Arkansans and Missourians left for Texas during the war, hoping to escape the war and retain ownership of their enslaved workers. The delay in constructing the trenches at Dooley's Ferry was in no small part due to the fear among Confederate officers that impressing slaves from local farms to do the work would cause wholesale depopulation of the area as white people sought to maintain their "property" (Magruder 1893). The impressing of slaves by the military for manual labor was a major cause of dissension among the Southern white population, and the estimates of up to 150,000 slaves that were taken into Texas during the war underscore the extent of the opposition (Baum 2008).

When Williams writes about using Dooley's Ferry to stop people leaving, he shows that Confederate authorities were using the site to restrict civilian movement and impose a level of control over the populace that had not been evident earlier in the war. That control accorded the Confederate military several advantages. It kept people in the area to farm and help provide for Confederate forage; it kept men who could be drafted into service within arms' reach; and significantly, it helped regulate and monitor the movement of civilians and supplies. Taken together with Special Orders No. 50, the troops manning the trenches at Dooley's Ferry were a powerful means of restricting the opposition's freedom to traverse the territory. Dooley's Ferry became a gate stopping Confederate civilians, not Union soldiers.

Returning to Joint Publication 3-24, *Counterinsurgency*, the implementation of "persistent surveillance systems" is essentially a means of panoptic population control. Today, militaries surveil with airplanes, satellites, and drones. During the American Civil War, monitoring the populace had to be done via boots and spurs. When Major General Magruder came to Arkansas in fall 1864, one of the first requests he made was to civil authorities for five companies of state troops to travel from Arkadelphia to Little Rock to gather up "stragglers, deserters, and persons liable to conscription" (Casdorph 1996:287). This kind of conscription served two purposes. First, Arkansas was always short of men for the Confederacy, so strong-arming resisters into the ranks was a source of needed manpower. Second, conscription provided control and placed men, particularly those with some service history, both under the gaze of military authorities and beyond the reach of US recruiters who would eagerly accept the services of disaffected former Confederates. Having Home Guard units circulating throughout the region also provided a police function. The regular forces sometimes provided the same form of control, as McPheeters's (2002) journal records, with groups of soldiers traveling out from camp and gathering up or otherwise contending with opposition.

Conclusion

The case for a protracted insurgent conflict in southwest Arkansas is based on a patchwork of small evidence, without a conclusive, definitive historical statement. Nevertheless, one would expect little else from a conflict that did not involve the bureaucratic machine of the US Army and that sought to operate by stealth and evasion. This can be the power of archaeological research on ephemeral aspects of conflict such as irregular warfare. Those activities that were intended to evade notice and reaction from major governmental authorities are never going to appear in the historical record with the same degree of documentation enjoyed by the conventional forces and their leaders. Historical archaeology has long prided itself on capturing some glimpse of the lives of those whom the chroniclers ignore. Here, it provides a means of glimpsing those who tried to avoid being chronicled.

Critically, the militarization of the landscape is not merely a defensive measure but can be easily turned into a means to control allied civilians and silence opposition in support of the aims of prolonging conflict. Even cursory study of the Confederate home front in Arkansas illustrates that the idea that the people were unified in defense of hearth and home is a piece of popular memory designed to bolster white senses of loyalty, superiority, and indomitability. It is a story deliberately constructed in the years since the war (Blight 2001). The South in general, and Arkansas in particular, was not a unified front against the North during the war, and attempts to claim the contrary are more about bolstering postbellum efforts to resolidify white rule than they are about the war itself.

By stepping away from historical approaches to conflict and embracing a comparative anthropological framework stretched across the landscape, we find a way to challenge the heritage aspects of the past and to break away some of the contextually specific data, which can be of use to both military planners who look for inspiration and insight and those who seek to criticize the impacts of warfare. It makes legible some of the more insidious and insular aspects of conflict, ones that we should remain conscious of, particularly in an era where hypermilitarism and jingoism are rife in many national cultures.

Acknowledgments

I thank Jamie Brandon for advising and codirecting much of the fieldwork at Dooley's Ferry. Tim Mulvihill also contributed, mapping large sections of the Dooley's Ferry entrenchments. Douglas D. Scott helped in the development of some of the ideas presented here, and Frederick Smith, Gray Gundaker, Matthew Liebmann, Marley Brown, and Scott Nelson helped foster them.

8
"Dirty Little Wars" in Northern Mexico and the American Southwest

Charles M. Haecker

An image portrayed in eyewitness accounts, novels, and cinema of the Spanish invasion of the Americas is of thousands of warriors, bedecked in feathered regalia, swarming around a few score of conquistadors. Bernal Díaz del Castillo, one of the adventurers who participated in Hernán Cortés's 1519–1521 conquest of the Mexica Empire, recalled one such encounter: "The Indians were charging us in such numbers that only by a miracle of swordplay were we able to drive them back and reform our ranks" (Díaz del Castillo 1956 [1632]:52). In later centuries this type of scenario morphs into one of soldiers of the British Empire surrounded by a Zulu army at Rorke's Drift, and by hordes of the Mahdist Revolt in the Sudan. On occasion, some US soldiers were in roughly similar situations albeit on a much smaller scale. Almost invariably the native warriors are vanquished by European and Euro-American soldiery. Notable exceptions include the rout of Cortés from Tenochtitlán (La Noche Triste), the annihilation of a British regiment at Isandlwana, and General George Custer's miscalculation at Little Bighorn.

In the case of Spain's invasion of the Americas, it is a long-held belief that the Spaniards' lopsided victories over Amerindians were inevitable. This belief is founded on Spain's obviously superior technology and tactics: 1) firearms and steel swords versus obsidian-edged spears and clubs; 2) steel armor versus quilted cotton armor; 3) use of the war horse and war dog as both tactical and psychological weapons; and 4) military formations that maximized Western-style weaponry effectiveness. The implicit corollary to this belief is that Amerindians were congenitally incapable of making appropriate modifications to their ways of waging war thus sealing their fate. This chapter, however, presents historical documentation and archaeological data that refute this concept. It gives recognition to those native tribes that not only modified their traditional methods of warfare but also introduced innovative techniques that could check, and even reverse for a while, alien encroachments into their homelands.

Asymmetric Warfare on New Spain's Northern Frontier

William Prescott stated that "the history of the Conquest of Mexico terminates with the surrender of the capital [of Tenochtitlán]" (Prescott 1992 [1843]:589). But Cortés in 1521 did not conquer Mexico; instead, he simply dismembered the Mexica Empire. Another twenty years would pass before Spain controlled Mesoamerica's major population centers. In fact, the conquest consisted of an unending series of armed expeditions in what is today northern Mexico and the American Southwest. Territorial expansion, however, was not sustainable without the steady influx of adventurers arriving from Europe. Rumors of easy and lucrative conquests in Mesoamerica and Peru encouraged thousands of men, many of them veterans of European and Mediterranean wars, to gamble their lives in the Americas. By the mid-sixteenth century, however, prospects dimmed for finding empires comparable to those subjugated by the previous generation of conquistadors.

Those adventurers not willing to settle down typically joined semi-professional militias, enticed by promises of eventual royal pensions, titles, and vast estates that included grants of indigenous labor and tribute. By the late sixteenth century there were hundreds of militias, consisting of Spanish Iberian, Creole, African-descended, and indigenous men-at-arms. Few realized the rewards offered by the Crown for their martial services; most died impoverished and forgotten. Taking defense into their own hands, militias conducted privately financed police actions against indigenous rebels, pirate strongholds, and fugitives fleeing from urban authorities. They also participated in punitive expeditions that sometimes reached well beyond the frontiers. These expeditions functioned as search-and-destroy missions that destabilized indigenous societies before their eventual absorption into New Spain (Lane and Johnson 2007; Schroeder 1998).

Conquest frontiers were deadly for both colonists and unconquered alike. As termed by Ferguson and Whitehead (1992:8–12), the frontier was a *tribal zone*, a place of regrouping for displaced Amerindians and other undesirables such as escaped African slaves and native apostates fleeing from the Inquisition. Within tribal zones there occurred an intermingling of ideas and material exchange to produce hybrid cultures or "new" tribes. One significant outcome of this hybridization was the development of a new approach to warfare that fused European and Amerindian warfare techniques.

The Spanish were the principal European innovators in developing new tactics and methods of warfare. They also adopted warfare techniques used by their European and North African adversaries (Jörgensen et al. 2005:12–23). It was not unusual for Spaniards who had fought the French, Italians, Dutch, and Ottoman Turks to try their luck as soldiers of fortune in New

Spain. In the Old World they fought as pikemen and harquebusiers in compact infantry units (*tercios*), as artillerymen, and as cavalrymen. They were led by generals who maneuvered armies that could number over 200,000. But Habsburg emperors could not afford the financial luxury and logistical nightmare of embarking even one troop company to New Spain.

Even if such a force appeared in New Spain, it would have offered little to the viceroy because it was trained for conventional, European-style warfare. By around 1540 warfare in New Spain had become very unconventional. The pike-bristling *tercio*, used with terrible effect when fighting on open ground, was a virtually useless formation within New Spain's jungles, mountains, swamps and deserts. Early firearms usually did not have a decisive advantage over indigenous weapons in terms of range, accuracy, or rate of fire; in wet climates they often did not work at all. Instead of quaking in fear from the sound of harquebus volleys and the unfamiliar sight of men on horseback, warriors might instead charge through the hail of shot and even run past the cavalry, while at the same time shooting arrows, swinging clubs, and tossing javelins at the astonished Spanish. The crossbow was obsolete in Europe around 1520, phased out by massed firepower of ranks of harquebusiers, yet in New Spain crossbows provided an adequate alternative when gunpowder became useless. Large-bore artillery fieldpieces were rare in frontier battles. Small-bore swivel guns, however, could decimate massed warriors.

During their European and North African wars, the Spanish adopted the tactic of mixing light cavalry (*jinete*) with infantry *tercios* to great effect. The *tercio* did not survive on New Spain's frontiers but the *jinete* was invaluable when pursuing an elusive foe that fought on foot. Frontier fighting required the Spanish militia to adopt Indian tactics. Most men-at-arms in New Spain adopted quilted cotton and layered leather body armor as it was effective and much cheaper to acquire than plate steel and chain mail armor. Search-and-destroy sorties consisted of experienced men who traveled deep into hostile territory. The attackers traveled on foot to minimize possible warning of their approach, with minimal equipment and supplies. They were on alert for possible ambush and could not expect reinforcements. Sometimes few or no attackers survived. The action was successful if hostile leaders were killed, entire communities obliterated, and captives taken for sale as slaves (Restall 2004).

A major setback to New Spain's territorial advances occurred during the 1541–1542 Mixtón War, wherein city-states of the Caxcan tribe, in the newly established province of Nueva Galicia, revolted against Spanish rule. During the revolt's initial phase, the Caxcan defeated a small army made up of about 100 Spanish men-at-arms and cavalry reinforced by 1,000 allied warriors. Thousands of Caxcan warriors then laid siege to the town of Guadalajara. Other tribes, sensing an opportunity to annihilate the hated Spaniards, joined the Caxcan cause. Even Mexico City was in danger of falling.

Viceroy Antonio de Mendoza scraped together another army, consisting of some 800 Spaniards and 60,000 Indian allies. This force lifted the siege then advanced toward Peñol de Nochistlán, a major Caxcan fortress city perched on a high, rocky hill (*peñol*), where the Caxcan had destroyed the previous Spanish army. To prepare for the attack, warriors constructed concentric rings of rock walls that encircled the hill's base, slopes, and escarpment. After three days of fighting, the overwhelming Spanish force reached the fortress city. Thousands of Caxcan fought to the death; many others killed themselves by jumping off a precipice. It was reported that "of the [Caxcan defenders] on the peñol, not even 20 escaped because those who were fleeing encountered the horsemen" (Muñoz 1541 in Haecker et al. 2007: 175). The Spanish army then laid waste to all the other Caxcan urban centers. By 1542 Spanish rule was reestablished, although conflicts in that region continued for decades.

The identification of Peñol de Nochistlán occurred in 2004 when archaeologists surveyed various prospective rocky hills within a 15 km (9.25 mi) radius centered on the present-day town of Nochistlán, Zacatecas (Haecker et al. 2007). In 2009 and 2010, archaeologist Angélica Enríquez conducted a survey of Peñol de Nochistlán (today named El Tuiche). This survey identified remains of the fortress city and three concentric rings of rock walls encircling the hill. Subsurface artifacts included crossbow bolt heads, harquebus shot, chain mail rings, equipage fittings, and obsidian arrow and atlatl dart points (Medrano 2012). Future surveys will investigate other Caxcan urban centers for the purpose of interpreting Spanish-Indian ally offensive tactics and Caxcan defensive measures. This project presents an opportunity to define a rare event in human history: the head-on meeting of Old World technology and tactics reflective of the early modern era with warriors of the New World who employed a radically different technology with its corresponding appropriate tactics.

Following the Mixtón War, the viceroy increased expenditures on recruiting and equipping salaried men-at-arms as region-wide rebellions could not be quashed solely by semiautonomous frontier militias. A new concern was the fighting strength of the Chichimeca nomadic tribes to the north of Nueva Galicia, who had fought the Spanish as Caxcan allies. But limited resources were overextended, thus expansion of New Spain's northern frontiers slowed. This situation changed in 1546, when prospectors discovered a silver-loaded mountain range some 600 km (960 mi) northwest of Mexico City. The place of discovery was called Zacatecas, where the most remarkable silver strike ever was made. Although New Spain was ill prepared to renew its northward advance, this discovery started a chain of events that changed the course of New Spain's economic development. For the next few years the Zacatecas prospecting venture shipped ever-increasing amounts of silver southward along a road network that led to Guadalajara and Mexico City. This road was the lifeline that provided mining camps with provisions including weapons. The weapons were needed because local native bands had become increasingly hostile. Lack of organized government in Zacatecas encouraged increased mistreatment of the local Chichimeca, and many were pressed into service as carriers or were enslaved to work in the mines.

By 1550 the Spanish realized too late that they had instigated a war not just with a minor Chichimeca tribe but with all four "nations" of La Gran Chichimeca. Chichimeca warriors made for an especially fearsome enemy numbering perhaps as many as 50,000 in the mid-sixteenth century. Their approximately 3,000,000 sq km (1,158,306 sq mi) homeland lacked an abundance of natural food resources. Their basic foods consisted of cactus fruit, seeds, and roots, supplemented with anything that moved. Their adaptation to limited food resources, however, gave them greater mobility than the sedentary invaders, who depended on domesticated livestock, agriculture, and imported supplies. Chichimeca encampments were hidden in caves, ravines, and narrow valleys. Rough terrain and thick vegetation cover typically surrounded an encampment as these features both obscured Chichimeca presence and negated the Spanish tactical advantage of cavalry.

The Chichimeca warrior's principal weapon was the bow and arrow. Warriors could release arrows with rapidity and accuracy at a far greater rate than a Spaniard could load and fire a harquebus. The force and penetrating power of the Chichimeca arrow were impressive; it found its way through multilayered leather armor and all but the closest-woven chain mail. Effective use of Spanish weapons was diminished by the suddenness of Chichimeca attacks at close range. The Chichimeca penchant for cannibalism and especially gruesome torture techniques added psychological warfare to their fearsome reputation. Knowledge of these possibilities at Chichimeca hands induced many Spaniards and their Indian allies to abandon frontier life, causing depopulation of mining camps and ranches, labor shortages, and difficulty in recruiting soldiers.

Initial Chichimeca attacks during the 1550s aimed at caravans, with hundreds of warriors engaged. With growing confidence, warriors later raided villages and ranches along the frontier. Usually their attacks occurred at dawn or dusk, a tactic learned from their Spanish adversary. The warriors' preferred method of attack was the ambush. Within a constricted area traversed by a road, warriors prepared attack positions by placing stacked rocks that blended into natural surroundings, then waited for their enemy. If pursued, warriors broke into smaller groups and dispersed over rough terrain. The Chichimeca acquired European weaponry through raids and by exchanging their booty of silver ingots with European traders. Many warriors mastered the use of swords and the harquebus. Most worrisome to the Spanish was that some Chichimeca tribes learned to ride horses instead of simply eating them. Mounted warriors conducted rapid attacks and withdrawals, and in some areas of their homeland, the Chichimeca possessed more horses than their adversary.

During the 1570s, the Spanish implemented measures for highway protection that included construction of wagon forts, which accompanied caravans. All miners and ranch workers were required to possess weapons and armor, ranches and settlements constructed blockhouses, and new settlements were strategically positioned with defense in mind. Occasional punitive campaigns continued; however, Spanish actions placed emphasis on defensive measures. Presidio troops primarily served as caravan escorts, and roving patrols roamed over the land in specified areas for defense and quick reprisal. These men-at-arms were recruited, paid, and trained by the viceroyalty. Frontier militias continued to serve as a stopgap when professional soldiers were unavailable.

Another policy shift included viceregal permission for Indian leaders of allied sedentary tribes to arm their warriors with swords, fight on horseback, and award these leaders with military commissions and gifts. It became essential that Indian allies bear the brunt of fighting in La Gran Chichimeca as there were not enough Spaniards to continue the war without their help. The greatest cause perpetuating the Chichimeca War was that Spanish soldiers provoked conflict by slave raiding peaceful Chichimeca communities, which in turn took vengeance for the raid. As a consequence, the viceroy forbade all unauthorized attacks, eliminated Chichimeca enslavement, and limited military activity to highway defense.

Assimilation of newly pacified tribes was accomplished through gifts of European goods and livestock. These actions had the desired effect, and by the end of the sixteenth century pacification of La Gran Chichimeca was complete. Frontier fighting, however, continued as the boundaries of New Spain advanced well beyond the northern reaches of La Gran Chichimeca (Bletzer 2016; Powell 1952). The northern colony of New Mexico, for example, was barely holding on during the mid-seventeenth century due to increasing attacks by various nomadic tribes that the Spanish collectively named Apache.

Apachería

The 1539–1542 Vázquez de Coronado expedition occasionally encountered Apache bands during its journeys through the American Southwest. Expedition accounts do not indicate loss of Spanish possessions by theft or death of expedition members as a result of these encounters. Archaeological evidence, however, indicates that at least some Apache bands acquired a few European-manufactured objects from the Vázquez de Coronado expedition. Sixteenth-century-dated Spanish metallic artifacts were recovered from Apache *ranchería* sites in southwestern New Mexico. These sites are located approximately 70 km (43 mi) from the closest segment of the expedition's likely route (Adams 2012). After Spaniards established the province of New Mexico in 1598, the Apaches substantially increased their food supply by taking livestock from Spanish farms, and over time they came to depend on raiding as a primary source for food. Raiding parties were small, usually between four and 12 men. Raiders initially traveled on foot, concealed themselves and watched their intended victims until the opportune moment, then stealthily drove off livestock. Apaches avoided fighting if possible, for it would jeopardize the purpose of the raid. Their strategy involved multiple warrior groups raiding widely dispersed ranches and settlements in New Mexico, Chihuahua, and Sonora at the same time, making it extremely difficult to organize an effective defense against them (Worcester 1979:4–19).

If any warriors were killed during a raid, the raiding band later returned for vengeance. First, small groups of warriors infiltrated the chosen area of operations. Each group obliterated any evidence of its presence and rendezvoused at a preselected base camp with a naturally defendable and camouflaged spot. Scouts then selected targets such as isolated ranches and undefended villages and determined a location for ambush where terrain would work against their enemies. The attack group quietly arrived at the chosen location before dawn. Apaches favored tactics based on surprise because it caused sudden fear and confusion in their foe. Surprise assault involved simultaneous attacks using archery, spears, and javelins. One such place of ambush was identified within Box Canyon in the Florida Mountains, New Mexico. At this location the Apaches constructed several blinds that blended into the surrounding rock formation. These blinds were positioned so that a warrior would be situated within 20 m (65 ft) from anyone traveling through the center of the canyon (Figure 8.1) (Haecker 2002).

During a fight Apaches might suddenly shift from offense to defense to trick their enemy into exposing himself. Irregular formations appeared to have separate objectives; however, this was a tactic of indirect approach, with several of these attacks being feigned. Apaches attempted to kill or capture

Figure 8.1. Breastwork (identified by arrow) within Box Canyon, Florida Mountains, New Mexico. Surveyor's head appears at the top of the breastwork. (Courtesy of Charles M. Haecker)

all enemies, then clear away signs of attack so that their presence would remain undetected. Lookouts communicated with the main party through smoke signals or flashing glass.

If an approaching enemy force was too strong, the Apaches broke into groups, dispersed in different directions, and withdrew to their *ranchería* inside their home territories (termed Apachería by the Spanish) (Basso 1971). Approaches to a *ranchería* were often screened by natural rock formations that could be easily defended. Sentinels were posted on nearby elevated formations overlooking the *ranchería*. Archaeological surveys designed to locate *ranchería* sites first identify these sentinel posts, then search for evidence of the site near the formation's base. Sentinel posts, positioned at the edge of an escarpment, are typically identified by a low rock stack that served to obscure the sentinel when in prone position (Figure 8.2) (Adams et al. 2000; Haecker 2002).

If an enemy force seemed likely to reach their *ranchería*, the Apaches usually abandoned it. Women, children, and elderly men left as a group. Warriors protected their flanks, with other warriors conducting a counterattack to slow or stop the invaders. Once out of immediate danger, the groups dispersed and later rejoined at some predetermined place. Deaths or capture

Figure 8.2. Sentinel breastwork overlooking an Apache wickiup village site, Last Chance Canyon, Guadalupe Mountains, New Mexico. (Courtesy of Charles M. Haecker)

of large numbers of defenders usually took place because of a surprise attack. Apache scouts and sentinels, however, typically reported an approaching enemy, and the majority of the *ranchería* population would successfully escape (Matson and Schroeder 1957:347).

Traditional weapons used by the Apaches were essentially similar to those of other Native American groups encountered by the Spanish in northern New Spain: spears, javelins, clubs, and powerful composite bows. Once they adopted the horse as part of their way of life, the Apaches used the lance in much the same way as did Spanish cavalry. By the mid-eighteenth century, captured swords and various other metal objects were often cut into pieces and reconfigured into lance heads and arrow points. Compared to Apache bows, the smoothbore muskets that replaced the less powerful harquebus by around 1620 were slow to load and inaccurate. Firsthand accounts indicate that an Apache could fire as many as ten arrows in the time it took to load and fire a musket. In contrast to arrows and javelins, however, a musket ball could cut through the thickest leather armor or shield. A musket ball striking flesh could produce a terrible wound that immediately stopped an opponent. Consequently, Apaches saw the value in acquiring firearms and mastering their use. By circa 1650 the horse had become an important part of the warriors' equipment: it was used as transportation for people and packing supplies and as food (Moorehead 1975:188; Schroeder 1983).

Another weapon used by Apaches and Spanish alike was calculated terror. Like the Chichimeca, Apaches tortured prisoners to exact vengeance and to demoralize their foe. Gruesome remains of those tortured to death were left in conspicuous locations so that the Spanish would discover them (Pfefferkorn 1989:148–149). And just as they did during the Chichimeca War, Spanish counterinsurgency tactics included, as an example, killing everyone in a *ranchería* to frighten their kinsmen. Invariably, this tactic had the opposite effect. Apaches in response attacked Spanish and native ally villages, committing essentially the same cruelties meted out to *rancherías* (Vargas Machuca 2008 [1599]).

By 1770 Spain's hold on its northern provinces was slipping. Apache raids, some now fielding hundreds of warriors, had destroyed numerous ranches and farmsteads, resulting in semistarvation in Spanish population centers. Colonists and peaceful Indians fled southward in fear of hunger and Apache rage. Even presidios were occasionally attacked, resulting in loss of presidio horse herds in addition to the deaths of presidiales. It was useless to pursue Apaches at one point only, for they were active along the entire frontier. In response, presidios were reorganized into a single line, so that each controlled an Apache raiding trail. Additionally, Spain made alliances with non-Apache tribes to wage a war of extermination against their now-common foe. No amount of strategic reorganization, however, could make up for the lack of sufficient troops and adequate arms and horses. Nonetheless, military reforms enacted in 1772 gradually improved the discipline and training of presidiales, and punitive expeditions became more effective. In 1781 the presidiales were issued improved firearms, which restored their morale and efficiency.

The Spanish in 1785 also initiated a more positive political framework with Apaches. They were allowed peace whenever they requested it. When peaceful, Apaches were furnished food supplies on a regular basis and encouraged to live near presidios where their movements could be observed. Pacification worked well up until the beginning of the Mexican War of Independence, in 1810. With its independence won in 1821, the newly formed Mexican government was unable to continue buying peace with the Apaches, whereupon the Apaches renewed their former raiding practices. Failure of the Spanish to conquer Apachería was not due to the Spanish lacking confidence in themselves; they had already subjugated other warlike peoples, such as the Chichimeca, during the expansion of New Spain's frontiers. Yet due to the Apaches' relatively low population numbers and wide dispersal of nomadic bands over a vast region, Spanish stratagems that worked elsewhere ultimately proved ineffective.

When, in 1848, the United States acquired Mexico's northern lands, it also found itself saddled with the "Apache problem." Battle-tested dragoons had never campaigned against Indians as elusive as the Apaches and, like their predecessors, were unprepared for the swift movements necessary to bring them to bay. Large, grain-fed cavalry mounts often could not keep up with wiry Apache ponies that thrived on desert scrub. On occasion a few warriors might tempt a troop into a mad chase across a desert flat. Troopers finally halted chase when their thirsty horses fatigued, then they backtracked to a known watering hole—where Apaches waited in ambush. United States frontier forces frequently underestimated the Apaches' mastery of asymmetric warfare tactics.

Battle of Cieneguilla (1854)

The 1854 Battle of Cieneguilla, fought in mountainous terrain south of Taos and near the village of Cieneguilla, New Mexico, was another example of underestimation of the Apaches' mastery of asymmetric warfare tactics. Captain John Davidson's 60-man command planned an early morning surprise of a Jicarilla Apache *ranchería* located on a ridgetop. The Apaches, however, were aware of the dragoons' approach and prepared a countersurprise. Upon reaching the base of the ridge, Davidson ordered his men to dismount. Their horses, guarded by 15 men, stayed at the ridge base while the larger detail advanced upslope toward the waiting Apaches. After a brief but fierce fight, the Apaches retreated down the opposite side of the ridge. The dragoons' self-congratulations ended, however, when they discovered that the horse guard was under attack. Forced to retreat, Davidson's men formed a defensive ring around their horses. For the next two hours the dragoons suffered aggressive attacks by over 100 warriors, who held high ground on three sides. The dragoons' inaccurate musketoons were inadequate against skilled bowmen and marksmen armed with trade rifles. Finally, the dragoons galloped through a gauntlet of arrows and regrouped at the base of a nearby ridge. They gained the ridgetop only to find warriors waiting for them. With ammunition now dangerously low, the dragoons scrambled down the opposite slope and escaped total destruction. They left behind the bodies of 22 of their comrades; 23 of the 38 survivors were wounded. The Battle of Cieneguilla was the most significant defeat of the US military in the West at that time (Johnson et al. 2009).

In 2001, an archaeology team identified the location of the Cieneguilla battlefield. The team recorded over 1,000 artifacts related to the battle and the Apache *ranchería*. Artifact distribution patterns provide insights regarding battle tactics employed by the dragoons and Apaches. The location of the *ranchería* is an indication that the Jicarilla had defense clearly in mind. Its placement on a ridge crest provided numerous escape routes in the event of an attack. Several nearby drainages also would have provided concealed positions for the Apaches when they counterattacked. Cordero's 1796-dated account of Apache tactics notes that "when the time came for attack, they frequently resorted to an ambush in a place favorable to them. This consisted in sending some of their fastest runners or mounted warriors ahead to draw out the enemy. When the latter attacked the swiftly retreating Apache, they suddenly found themselves surrounded by the main body and cut to pieces. After the battle, the Apache retreated with incredible speed, deliberately following a rout through rough, mountainous land to discourage pursuers" (Matson and Schroeder 1957:345–346; Cordero cited in Johnson et al. 2009:66–67).

These are the tactics used by the Jicarilla at Cieneguilla. Artifact patterning indicated various Jicarilla tactics that included pinning down the dragoons and then making sporadic frontal attacks while shooting from behind boulders and trees. This action forced the dragoons to retreat upslope to the crest of another ridge, where other Jicarilla pinned them down on a narrow neck of the ridge, then sealed them in on three sides. In this final phase of the battle there are more arrow points than Jicarilla-fired rifle balls, suggesting that the Jicarilla either had run out of firearm ammunition or had saved the arrows for close-quarters fighting where these projectiles would be more effective. Archaeological evidence indicates that the dragoons used their standard tactic of approaching the *ranchería* in full view, then dismounting before their upslope attack. Based on the outcome of the battle, Davidson's choice of locations for dismounting was a poor one in that he did not consider the possibility of having to defend a position if the Jicarilla should counterattack while holding high ground, which they did.

Battle of Apache Pass (1862)

On the eve of the Civil War, the Apache conflict in Arizona had reached a crisis. Army posts were too widely scattered and poorly manned to pose a significant threat to raiding Apaches, who roamed virtually unrestricted over the entire region. Finally, by spring 1861, the withdrawal of federal troops from Arizona to eastern theaters of the Civil War presented Apache chiefs Mangas Coloradas and Cochise an opportunity to annihilate all remaining intruders. They assembled an arsenal of rifled muskets, shotguns, and pistols and strengthened alliances with other Apache bands. In early July 1862, warriors prepared Apache Pass in southeastern Arizona as a place of ambush against US troop formations that funneled through it. The chiefs directed construction of breastworks having overlapping fields of fire and overlooking the canyon's only spring. At the western approach to the pass they positioned warriors whose duty was to prevent soldiers from escaping once the attack began. If their tactic worked, all trapped soldiers would die either from gunfire or from thirst.

Around noon on July 15, 1862, a 96-man company of New Mexico–bound California Volunteers entered the western approach into Apache Pass. After marching for 12 hours the exhausted soldiers needed water from a spring located deep within this pass. Lying in wait were scores of Apaches led by Chiefs Mangas Coloradas and Cochise. The chiefs placed their men in two advantageous positions. One group hid on both sides of the western approach, in an area of rugged terrain. Here the soldiers would be forced to proceed slowly, making them easy targets. A larger group of Apaches was positioned on high ground overlooking the spring, hidden behind rock-stacked breastworks, boulder outcrops, and thick underbrush. Once the soldiers reached this location they would be exposed to withering fire from both sides.

The plan went awry when some warriors fired their weapons at the company's rear guard before it entered the pass. This group of 20 men was escorting wagons and two 12-pounder howitzers. The rear guard fended off this attack then hastened to catch up with their comrades. Distant gunfire alerted company commander Captain Thomas Roberts, and it also signaled to Apaches positioned within the pass to open fire. Roberts ordered half his men to advance toward the spring in skirmish formation. The others, led by Roberts, fought their way upslope and drove back the Apaches. By then the rear guard arrived and immediately wheeled the two howitzers into action, firing several rounds of spherical case shot in the warriors' general direction.

By late afternoon the soldiers commanded the spring and quenched their thirst. Roberts determined that his outnumbered men could not hold both the spring and the surrounding high ground, so they retreated onto relatively less constricted and thus more defensible terrain west of the spring. The Apaches soon returned and reclaimed previously abandoned positions. Any soldiers who tried to gain access to the spring were driven back by sharpshooters. The grim situation of Roberts' command improved the next morning when reinforcements arrived. Daylight once again permitted use of howitzers, which were advanced down the pass and were periodically fired using shrapnel-scattering case shot at various elevated positions believed held by the Apaches. Finally, soldiers charged up the slopes—to discover that the Apaches had melted away (Fountain 1962:33). The two chiefs presumably realized that success was now impossible, and no value remained in continuing the fight.

The soldiers' retelling of the battle turned the event into a struggle of epic proportions. Newspapers picked up the story and interviewed several soldiers, who believed that there were as many as 800 Apaches, all of them armed with state-of-the-art Sharps carbines, rifled muskets, and revolvers. Captain Roberts reported two soldiers killed and nine wounded. He could verify six dead Apaches but believed his men killed many more. Albert Fountain (1962:32), a veteran of the battle, decades later recalled that "For hours, the air was constantly filled with lead," and "the shells burst splendidly." Conversely, Apache versions understate the incident, claiming that the soldiers killed few warriors (Ball 1970:46). Were these accounts true? The actual location of the battle was not in question; rather, battle intensity and complexity were open for testable questions: What varieties of firearms did the Apaches use? Did the Apaches construct breastworks? Approximately how many Apaches participated in the battle? Was artillery fire as intense and accurate as described by the soldiers?

Using metal detectors, archaeologists swept high ground overlooking the spring, the bottom of the pass where soldiers held defensive positions, and the western approach to Apache Pass where the Apaches attacked the rear guard. Resultant data produced an artifact distribution map that resurrected the Battle of Apache Pass. Of 362 artifacts recovered, 84 percent were found on high ground overlooking the spring. In addition to shrapnel, this artifact grouping included fired .58 caliber Minié balls, .52 caliber Sharps carbine bullets, and .36 caliber pistol balls, which reflect all known types of firearms issued to the California Volunteers. Additionally, these bullets tended to cluster around a rock-stacked breastwork that overlooks the spring, indicating the soldiers' attempts to neutralize this Apache position.

Within areas where soldiers once held their positions, archaeologists discovered Apache-fired bullets derived from a mix of military issue and nonmilitary firearms. The Apache-fired bullets represented only 4 percent of the total artifact count. Fragments of exploded cannonball shell and lead ball shrapnel were scattered primarily on high ground overlooking the spring; however, evidence of case shot was found about 2 km (1.2 mi) away from Apache Pass. Survey of the vicinity where the rear guard was attacked produced 12 percent of the total artifact count.

Together, these findings indicate that most of the battle took place in the vicinity of the spring. The soldiers' firepower was superior to that of the Apaches, suggesting that the latter neither significantly outnumbered the soldiers nor were uniformly armed with quality weapons. There is only limited protective cover within the few hectares of high ground that confines the spring, making it unlikely that many more than 100 warriors could safely hide themselves there. In fact, the Apaches did not require superiority in numbers. All they had to do was expend enough ammunition to keep the soldiers both pinned down and trapped away from the spring. The amount of shrapnel recovered on high ground was surprisingly sparse—just two specimens of spherical case shot could have accounted for all recovered shrapnel balls found there. This discovery, combined with widespread dispersion of shrapnel well beyond the confines of Apache Pass, suggests that the howitzers were fired far less often and with less accuracy than what was later recalled by the soldiers (Haecker 2001; Ludwig 2016; McChristian 2005:48–61).

Victorio's War (1879–1880)

The cause of Victorio's War was the US government's 1876–1877 policy that concentrated mutually hostile Apache tribes onto the San Carlos Reservation in southeastern Arizona. Enactment of this policy meant closure of both the Chiricahua Apache Reservation at Fort Bowie, Arizona, and the Warm Springs Apache Reservation at Ojo Caliente, New Mexico. San Carlos was placed within a desolate and unhealthy region that lacked natural resources needed to support concentrations of humans. A dysfunctional and corrupt Indian Agency also failed to provide basic subsistence for the Apache. The result was reservation-wide starvation and disease.

Victorio, a warrior and chief of the Warm Springs band of the Tchihende division of the central Apaches, maintained his claim that Warm Springs was his people's rightful homeland, and from 1877 to 1879 he made several attempts to either negotiate or fight for their return to Ojo Caliente. By August 1879, Victorio concluded that the US government would never permit them to leave San Carlos. Consequently, he led several hundred men, women, and children back toward their Ojo Caliente homeland located over 170 km (105 mi) away. Other disaffected Apaches eventually joined Victorio's band that at most included no more than 200 warriors.

Colonel Edward Hatch, commander of the 9th Cavalry Regiment, responded to Victorio's breakout with the classic counterinsurgency strategy of hot pursuit. But Hatch was burdened with having to assemble an ad hoc pursuit force from a widespread scattering of forts and outposts. As a consequence, the resultant force of approximately 600 men lacked overall unit cohesion and, at the same time, its formation stripped military protection from frontier settlements. To work well, Hatch's strategy required well-coordinated companies to move rapidly, make contact with the hostiles as soon as possible, then surround and overwhelm them using superior firepower and numbers. This strategy was flawed, however, due to its dependence on forward supply bases, lack of remounts, and chronic miscommunication between companies. For the next seven months Victorio exploited these and other military weaknesses. For example, the Apache leader avoided easily traversed and well-watered routes; instead, he led his band over rough mountain and desert terrain while purposely leaving signs or human decoys for his pursuers to follow. Large numbers of irreplaceable cavalry mounts consequently became lame or died from exhaustion; thus many a trooper finished the campaign on foot. Apaches likewise lost mounts in large numbers, but for them a dead horse meant dinner. They readily obtained remounts by raiding isolated ranches and communities, even capturing horses from army supply bases. Victorio also traded unwanted loot with unscrupulous traders in exchange for firearms and ammunition.

One cavalry force that pursued Victorio into Mexico discovered that the Apaches had churned up desperately needed watering holes into undrinkable mud bogs, with putrid animal carcasses tossed in for good measure. Weakened by lack of mounts, desperate for water and with morale sagging, pursuer now became prey. By slowing down his pursuers, Victorio purchased time to select and prepare places for ambush. If the cavalry commander continued pursuit or instead chose discretion and return to his supply base, results would be the same: he and his men would be surprised by withering fire from multiple directions. The Apaches first aimed at the troopers' horses and pack mules because these were larger targets. More horses and mules killed made it less likely that the cavalry company could extricate itself from a life-threatening situation.

Victorio adhered to the Apache maxim of doing maximum harm with minimum loss. Consequently, when cavalry reinforcements finally arrived, he broke off the attack. Some of the warriors shielded the band's women and children as they quickly exited out of harm's way. A remaining screen of warriors gradually disengaged from the fight as they retreated upslope to maintain supremacy of high ground. Victorio never managed to annihilate an entire cavalry company, primarily because officers maintained command and control, and their men fought well. Also, each trooper knew his fate if he broke and ran: either death by a hail of Apache-fired bullets or capture only to be tortured to death (Thrapp 1980).

Battle of Hembrillo Basin (1880)

In the late afternoon of April 6, 1880, two companies totaling 71 men advanced toward Victorio's *ranchería*, positioned on high ground overlooking

Hembrillo Basin. Awaiting them were approximately 150 warriors, placed along two low ridges that form a V-shaped constriction within the basin. They opened fire only when both companies had entered the constricted area of the basin. The troopers quickly dismounted with every fourth man holding the reins of horses and pack mules, thus dropping their skirmish line strength to 50 troopers. Victorio possessed numeric superiority, held the high ground, and controlled access to a spring-fed stream, the only water source within the basin. The troopers formed two skirmish lines with their backs to each other, in response to the V-shaped trap that they had entered. Their position was perilous as the backs of both lines of skirmishers were exposed to rifle fire. Horses and mules used for cover were prime targets.

Nightfall provided some respite for the troopers; however, lack of water now became their main concern. Attempts to reach water located about 300 m (984 ft) away were foiled by warriors who guarded the approach to it. Other warriors crept ever closer toward the troopers, who were now pinned down on all sides. Throughout the night some Apaches cried out blood-curdling war chants; others who spoke English shouted out detailed descriptions as to what they would do to captured troopers. By daylight, 25 army horses and pack mules had either been killed or had stampeded and been captured by the Apaches. But daylight also saw the arrival of two reinforcement companies that initiated the dangerous process of extracting the other two companies.

Faced with a now numerically superior enemy, Victorio first removed to safety the band's women and children in the *ranchería*. High ground on the battlefield was still held by a gradually retreating screen of warriors. Their rifle fire discouraged troopers from obtaining water and reinforcement companies from conducting flanking maneuvers. Finally, cover fire from one group of troopers permitted another group to rush the Apache position. On reaching the top of the ridgeline, the troopers discovered that the Apaches' position, which had been so hotly contested a few minutes before, was abandoned. Once again, Victorio's band had slipped away.

Karl W. Laumbach of Human Systems Research conducted archaeological research of the Hembrillo Basin battlefield. His survey identified a broad pattern of breastworks that Victorio positioned to augment the natural fortifications of limestone-capped ridges that enclosed the basin. Laumbach determined that some of these breastworks were in tactical locations that did not play a part in the actual battle and concluded that most of the breastworks must have been in place long before the battle began. Stated another way, Victorio not only chose the place of battle that was most advantageous for the Apaches, he also made preparations that took into consideration various scenarios regarding how the battle might progress.

The survey recovered over 1,100 fight-related artifacts, of which over 800 are cartridge cases. Archaeologist Douglas D. Scott conducted firearms identification analysis of the cartridge cases and identified 145 different rifles and carbines and 37 different pistols. Using a GIS (Geographical Information System) mapping program, it was possible to recreate movements of specific Apache and trooper firearms, and possibly combatants, across the battlefield. Cartridge case distribution patterns in relation to breastwork placement also provides insight into Victorio's tactical sophistication. The troopers were armed with Model 1873 carbines and Colt or Smith & Wesson .45 caliber pistols. The Apaches also possessed these firearm models as well as a variety of obsolete military and state-of-the-art, nonmilitary models.

Cartridge case distribution patterns indicate that 13 warriors, who were exclusively armed with Winchester and Henry repeating rifles, formed one battery that zeroed in on the water source. Those few troopers who tried to fill their canteens during the night were driven back by a sheet of rapidly fired bullets. Conversely, warriors who fired at the troopers' skirmish lines were armed exclusively with military carbines and rifles. These powerful firearms could reach enemy positions located as much as 400 m (1,312 ft) away. Although accuracy at that extreme range was unlikely under these circumstances, accuracy was not a necessity. Rather, the Apaches' potshots served to kill, wound, or stampede horses and pack mules, as well as pin down the troopers. The troopers also left physical remains of their tactical response in the form of dropped cartridges, cartridge cases, personal items, and pieces of equipage. These artifacts denote the various positions held by the troopers, such as skirmish lines formed along the edge of a dry drainage bed. As a whole, these positions are poorly sighted for both offensive and defensive actions and reveal a force that could only react to aggressive tactics borne out by their enemy.

The Hembrillo Basin battle was an exception to the typical Apache method of warfare in that Victorio selected a battle area having wide-open terrain. Because of the location's openness, it is possible that the cavalry was not expecting an ambush when they first entered the basin, so their guard was down. This made them better targets. And because Victorio possessed powerful military carbines and rifles, he was able to spring an ambush at long range, something that had never been done before, or after, by any Apache leader (Laumbach 2001).

Victorio's ability to interweave Apache ways of waging war with advanced American weaponry, and his complete control over the place of battle while exploiting flaws in the US military's tactics and operations, makes Victorio's War a valued source of information for the US Army Center of Military History. The center has organized staff rides at several Apache War battle sites including Apache Pass and Hembrillo Basin as a means of impressing on young officers the basic dos and don'ts of engagement in asymmetric warfare at the tactical level (Saunders 1989:43–44). Archaeologists who investigate Apache War battle sites assist in these staff rides by pointing out the often-subtle physical evidence of Apache-prepared ambush localities and both Apache and US military defensive positions (Karl Laumbach, personal communication; Larry Ludwig, personal communication 2016).

Hembrillo Basin proved to be Victorio's last major fight against the US military. Six months later Mexican forces killed him and most of his band at Tres Castillos. By 1890 hostile Apache actions in both the United States and Mexico had ceased.

Conclusion

We identify the plethora of today's Latin American insurgency movements that typically employ guerilla warfare tactics as the progeny of a centuries-old, New World martial tradition. If one were to conduct archaeological investigations of guerilla-prepared ambush sites associated with, for example, the 1950s Cuban Revolution, physical evidence may be obtained that is strikingly similar to what is found on comparable Apache War sites. We identify Apachería as an ideal region for investigating the culmination of centuries of warrior ingenuity, which more often than not confounded far more technologically advanced adversaries. Finally, we recommend that the former homeland strongholds of La Gran Chichimeca become a focus of conflict archaeology studies. Here one might best identify the nascent attributes and rapidly evolving nature of asymmetric warfare as conducted in northern Mexico and the American Southwest.

9
The Hatfield-McCoy Feud as Asymmetric Warfare

Archaeology at the Randall and Sally McCoy Homestead

KIM A. MCBRIDE

Anthropologist Keith F. Otterbein defines feuding as a series of revenge killings occurring between kin groups within a political community (Otterbein 2009:44, 114). Otterbein's investigations shed light on kin-based, localized, low-intensity irregular warfare with cultural roots tracing back to Scots-Irish clan warfare (Fischer 1989; Otterbein 2000:231). One of Otterbein's case studies is the Hatfield-McCoy feud, one of the more iconic feuds in American history, which took place in the late nineteenth century within the Tug Fork of the Big Sandy River. The Big Sandy defined a part of the border between Kentucky and West Virginia and so both state governments were involved. The Tug Fork was in the late nineteenth century a land with vast timber and coal resources, which would play a role in the 1888 raid on the Randall and Sally McCoy homestead (Figure 9.1). This chapter presents archaeological investigations conducted in 2012 and 2014 at the site of this raid, and it places the site and raid within the interpretive context of irregular warfare.

The Hatfield-McCoy feud was a complex series of violent encounters and has been covered extensively in recent media such as a History Channel miniseries and a growing body of print sources (Alther 2012; Dotson 2014, 2015a, 2015b; Hatfield and Spence 2003; Hill 1888; Jones 1948; King 2013; McCoy 2015; McCoy 1976; Rice 1982; Spears 2015 [1888]; Staton 1993; Waller 1988). Otterbein (2000) analyzed the Hatfield-McCoy feud and four other feuds that took place in eastern Kentucky (Turner-Howard feud of Harlan County, French-Eversole feud of Perry County, Martin-Tolliver feud of Rowan County, and Baker-Howard feud of Clay County). He asserted that these events met the anthropological definition of a feud as each included the involvement of three or more killings or acts of violence by kin groups,

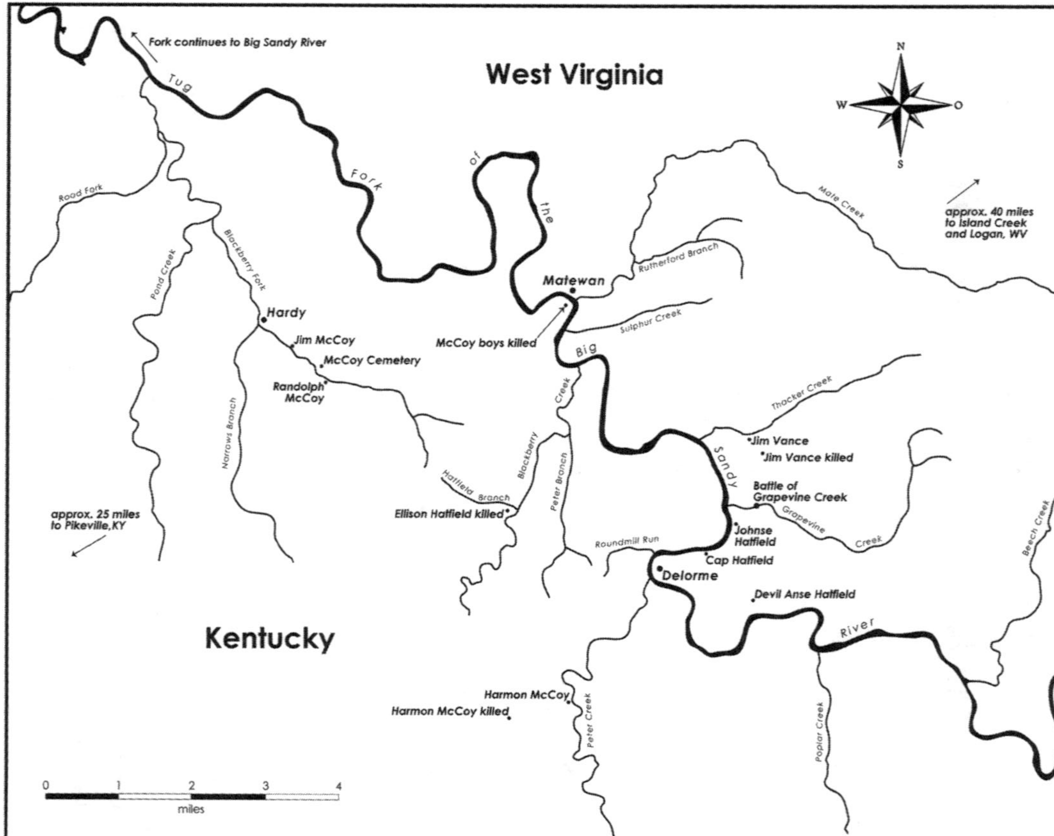

Figure 9.1. Hatfield-McCoy feud area and key sites mentioned. (Courtesy of Kim A. McBride)

often as revenge for perceived injustice, and occurred within an overall political authority (Otterbein 2000:231–232). Otterbein noted the preadaptation of eastern Kentucky residents to feuding, given their high marksmanship that rewarded an ambush style of fighting and their common heritage from the English/Scottish borderland where raiding and feuding was common. Typical factors in Kentucky feuds were pride, honor, politics, loyalty to family and friends, and whiskey (Otterbein 2000:241; Pearce 1994).

Otterbein identified seven categories of feud tactics or homicidal encounters that can be used to better understand the eastern Kentucky feuds: ambush, arranged battle, assault with knives, chance battle, gun fight, house attack, and murder of prisoners. Whereas the other four eastern Kentucky feuds in his sample had from two to five of these encounters, the Hatfield-McCoy feud exhibited six of them, lacking only an arranged battle. It also had a low percentage of encounters resulting in death, only 20 percent, compared with 44 percent of encounters in the larger set of feuds he studied (Otterbein 2000:239).

Of special interest is the form of encounter relevant to the Randall and Sally McCoy event: a house attack. Such attacks make up only 14 percent of all encounters in the Otterbein sample, the most common being ambushes, which make up 44 percent of encounters. Also, most of the other house attacks that are documented were more of an attack "at" a house, and not so much a total attack on a property resulting in its complete destruction. At the Randall and Sally McCoy homestead, the devastation was so complete that the two chimneys were the only things standing, and the site was subsequently abandoned by the family. The only comparable feud encounter in the sample was the 1885 burning of the Martin house in the Martin-Tolliver feud (Otterbein 2000; Johnson 1916). This total assault on a house strategy is more typical of earlier Civil War guerilla warfare than of a late nineteenth-century feud encounter. The attack on the Randall and Sally McCoy homestead resembled a Civil War–era raid because the participants were former combatants.

The January 1, 1888, nighttime Hatfield raid involved from nine to 14 (accounts vary) Hatfield family members or supporters (hereafter the Hatfield gang) mobilized to attack the Randall and Sally McCoy homestead on Blackberry Fork of Pond Creek in eastern Pike County, Kentucky, near the present-day community of Hardy (see Figure 9.1). Hatfield leader Anderson "Devil Anse" Hatfield was not at this raid, possibly due to illness, but also possibly due to his questioning the wisdom of the undertaking (Hatfield and Spence 2003). Supposedly, the party had attempted the attack the previous night but retreated after losing the element of surprise by knocking over a split-rail fence on approach to the house (Hatfield and Spence 2003).

Historian Otis K. Rice (1982) identified the Hatfield gang as including Tom Mitchell (sometimes Chambers), French or Doc Ellis, Charles Gillespie, Cap (also known as Little Cap) Hatfield, Johnse Hatfield, Bob Hatfield, Elliott Hatfield, Ellison Mounts (sometimes called Ellison Hatfield, as he was the son of Devil Anse Hatfield's brother Ellison Hatfield Sr.), and James (Uncle Jim) Vance (Devil Anse's uncle on his mother's side). James Vance and Cap Hatfield were the leaders. This list matches that given by Ellison Mounts, the only person convicted (and hanged) for the attack, in his confession. When Kentucky governor Simon Buckner wrote to West Virginia governor Emanuel Willis "E. W." Wilson on January 30, 1888, to complain of the raid, he described the group as a "company of twelve men, headed by Cap Hatfield and James Vance, Sr." (Dotson 2014:292). Note the reference to a military organization of a company in this description. This, and other accounts that give a higher number for the Hatfield gang, do not list the participants.

There are many conflicting accounts of the raid published in the various books on feuds, but this discussion relies on three accounts: the confession of Ellison Mounts (1888), who was present on the raid; John Spears (2015 [1888]), the first historian to give a lengthy report of the feud; and Coleman A. Hatfield, grandson of Devil Anse Hatfield, as reported by Coleman C. Hatfield and Robert Y. Spence (2003). The accounts are consistent in stating that at the time of the raid, Randall, Sally, son Calvin, and grandson Melvin were in the main house, and daughters Alifair, Addie, and Fanny, and granddaughter Cora were in the kitchen.

Tom E. Dotson (2014) argues that Mounts's confession was suspect as it was probably heavily coached by lawyers. He argues that the role of James Vance in particular was exaggerated. Vance was killed near his home on Thacker Creek, Logan County, West Virginia (see Figure 9.1) on January 7, 1888, by a Kentucky posse that included members of the Randall and Asa Harmon McCoy families. As a result, several members of the posse were under indictment in West Virginia for Vance's murder. Given this situation, if the lawyers could make Vance's involvement in the raid look worse in any testimonies that were given, this would be better for the defense.

Mounts's testimony is summarized as follows. This description starts at about 10:30 p.m. on January 1:

> Cap Hatfield and Jim Vance put on false faces as disguises and then Vance positions various members to go to the upper and lower doors of the kitchen and "Big House" [separate pens connected by a breezeway, see below] and orders that if any men come out, even in women's clothing, to capture them. Jim Vance disguises his voice and tells

> the McCoys to come out and surrender as *prisoners of war* [emphasis in original]; the McCoys do not respond. Cap Hatfield says if they will not come the house should be set on fire. Vance lights cotton and place[es] it under the shutters and in a joist hole. All parties shoot a number of bullets through the two doors of the big house and the McCoys inside the house return fire. Calvin McCoy [son of Randall and Sally] goes to the loft and shoots at the Hatfields. Calvin McCoy calls to his sisters Alifair, Fanny, and Addie, who are in the kitchen, to come put out the fire. Jim Vance tells the gang that if anyone comes out of the kitchen to shoot them. Cap Hatfield reminds them all to obey their Captain, Vance being known as their Captain. A girl [Alifair] comes to the door and Cap tells Mounts to shoot her. Mounts shoots Alifair with a .32 W.C. [Winchester] Rifle and she falls dying to the floor. Calvin McCoy is in the loft shooting at every opportunity, and hits Mounts in the left arm. Aunt Sally McCoy comes out and asks them to help put out the fire, save their belongings, and spare her life. Johnse Hatfield later tells Mounts that he struck Sally with a rock and a gun and left her for dead, but Mounts did not see this. Calvin McCoy broke from the house and ran in the direction of the corn crib and was followed by a volley of shots until he fell dead [later Mounts says they did not know that night that they had killed Calvin]. Cap Hatfield and Johnse Hatfield both claimed the shot but Mounts thinks from the position of the two that Cap Hatfield killed him. Randall McCoy breaks from the house with a double barrel shotgun loaded with birdshot, shoots Tom Mitchell in the right hand and some of the shot went into Mitchell's belt, then Randall shot Johnse Hatfield in the top of the right shoulder while Johnse was shooting at Calvin McCoy. Randall McCoy escapes into the woods. In a few minutes the raiders leave, the house is in "good flame," the women crying for help.

Historian John Spears (2015 [1888]) published a detailed account of the attack and the entire feud shortly after it took place. Although he only interviewed McCoy family members or individuals who were their allies, Spears's account probably has similar biases about the role of Jim Vance to those cited for Ellison Mounts's confession. The annotations by Ryan Hardesty (2015) in a recent publication of Spears's history counter some of the errors. Spears's account is less detailed but similar to that of Mounts's, differing in these points:

> The Hatfield gang (still headed by Jim Vance) call out to Randall to surrender or they would burn his house down. Charles Gillespie and

Elliott Hatfield were posted up and down the road while the other seven attacked. When Randall does not respond they try to break the door of the house but it is barred. Then the Hatfield gang breaks the kitchen door. Alifair tells them there are no men in the kitchen, and Jim Vance orders her to "stir a light." She goes to the fireplace and tries but fails. Cap Hatfield also tells her to strike a light or he will shoot her. Cap raises his gun to shoot but Ellison Mounts already has his gun at his shoulder and shoots her, she soon dies. One of the gang sets fire to cotton found sitting between the house and kitchen and throws it into the kitchen, using its light to see that there are no men in the kitchen, at which point all attention turns to the main house, where some of the Hatfield gang are already engaged shooting the southerly [back?] door to pieces. Someone puts burning cotton against the door, Randall takes buttermilk that is sitting in a churn and tries to put out the fire with it while holding a revolver with his other hand. Tom Mitchell climbs up with a flaming torch to burn the roof, holding it against the clapboard; Randall shoots his hand with a revolver. Calvin McCoy, who is in the loft, kicks off the gable boards and some clapboards and roof boards to shoot at the gang, hitting Ellison Mounts in the arm with a .32 caliber rifle, and driving the gang under the porch on the south [back?] side of his end of the building and between the buildings. The Hatfield gang keeps up fire at the door from their secured location under the porch roof, and continues to apply torches to the house roof while others fired the roof of the kitchen. Sally McCoy comes out the door facing the road [front, north] and goes around to the door of the kitchen toward Alifair and Jim Vance orders her back and hits her in the ribs with his gun, then Johnse Hatfield hits her in the head with his revolver. Calvin McCoy is heard telling his father Randall that he will try to reach the corn crib, which is about one hundred yards away across the yard to the northwest. Calvin runs out with cartridges in one hand and rifle in the other, the gunfire at him is quite loud and when he is about 75 yards from the house a bullet hits him in the head though the Hatfields do not know he is dead. Randall McCoy soon follows into the yard, shoots Johnse Hatfield in the right shoulder with a double barrel shotgun and shoots Tom Mitchell in the belt buckle creating a slight wound to his abdomen. Randall reaches the corn crib and escapes up the mountain on the other side of Blackberry Fork, finding a neighbor's barn where hogs had been sleeping to take shelter for the night. As Calvin leaves the house the gang spreads out from their retreat at the west end of the house. The roof of the kitchen caves in and Addie and Fanny [and probably Cora] leave the kitchen. Seeing the Hatfield gang then leaving on the road to the Tug Fork, they pick up the dead body of Alifair and drag it to the edge of the woods southeast of the house and built a fire, followed by Cora (daughter of Randall and Sally's son Tolbert, killed by Hatfields in 1882 in retaliation for his [along with his brothers Pharmer and Randall Jr. often called Bud] killing of Ellison Hatfield [Sr.]); and mother Sally crawls to join them. Melvin McCoy, another child of the deceased Tolbert McCoy, tried to leave the house clutching Randall's leg but trips and falls at the door, and crawled away around another corner of the house, all remain by the fire built by the girls all night.

We can get a Hatfield perspective from the writing of Devil Anse's grandson, Coleman A. Hatfield, as published by his son Coleman C. Hatfield and Robert Y. Spence in 2003 (167–170). The most important differences or perspectives in this account are as follows:

The idea of the raid comes from Elias Hatfield, partly because he is very concerned that McCoys are watching his house and planning to attack it. Devil Anse's support is lukewarm but other family members support it and Jim Vance agrees to lead it. Devil Anse draws upon his Civil War experience in planning the raid. They make a failed attempt on December 31, 1887 but turn back after dislodging a fence and making much noise, returning the next night with the intention of kidnapping Randall McCoy. Jim Vance approaches the door of the house yelling "Stir us up a light" at which point they hear Calvin McCoy loading his new .38 caliber Winchester rifle. Calvin McCoy shoots through the door just as Jim Vance jumps aside. Vance has an old muzzle-loader shotgun in his hand that goes off and sends a bullet through the porch shed where a bag of cotton is stored, catching the cotton on fire. Tom Mitchell climbed up on the porch roof in the back where there were windows, and Randall grabbed a shotgun and shoots Mitchell's hand [all accounts agree Mitchell lost three fingers]. Calvin is heard to say, "Pap, I'm going first. They'll kill me but let me go first, and you get away," and runs out and is shot. Alifair ran out of the house with some water to try to put out the fire, but as her water ran out she went back for a churn or pail of milk; "It is my understanding that there was no intention to kill the women. But when Alifair dashed out with that milk, Ellison Mounts, thinking it was a man, shot her. There has been some talk that Cap fired the shot, but Ellison Mounts confessed on the scaffold when he was hanged at Pikeville in 1890." . . .

The attack happened in less than a minute after Jim Vance jumped

> out of the way of the shot that came out from the McCoy house and fired the shot that set the cotton on fire, Jim Vance and Johnse Hatfield piled kindling against the house to start a second fire. Sarah (Sally) McCoy came out and "Jim Vance or maybe another clubbed her with a gun." (Hatfield and Spence 2003:170)

On January 2, the surviving McCoy family members relocated to Pikeville, Kentucky, the county seat of Pike County. Among those who moved to Pikeville were Randall's adult sons Jim and Sam, who abandoned homes near Randall's in the Blackberry Fork Drainage. The Randall McCoy family moved in with their cousin Perry Cline. Cline was sheriff or jailor of Pike County at one time. He had played a large part in earlier litigations of the feud and in encouraging raiding into West Virginia to capture Hatfields and their allies. Another important ally of the McCoys in Pikeville was Perry Cline's guardian, Colonel John S. Dils. Dils had been a colonel of the Unionist Home Guard, in which position he was often at odds with Confederate guerilla Devil Anse Hatfield. By 1888 Dils was one of the most wealthy, powerful, and politically connected men in Pike County (Dotson 2014; Hatfield and Spence 2003; Waller 1988). Dils gave Randall McCoy a job operating Dils' Ferry across the Cumberland River (Dotson 2014; Jones 1948; McCoy 1976; Rice 1982).

Historical Background of the Randall and Sally McCoy Site

To provide historical context for the site and the domestic life of the Randall (formally Randolph) and Sally McCoy family, a variety of primary records were investigated, including deeds, tax records, and federal census records (McBride and Shipp 2015). These primary records showed that Randall and Sally McCoy had lived at the site since at least the mid-1850s when they inherited between 50 and 100 acres from Sally's father (and Randall's uncle), Samuel McCoy. By the 1880s they added at least 400 additional acres to their farm. Sally and Randall McCoy had 16 children. In all the years of census enumerations, the household consisted of Randall, Sally, some combination of whatever children were still at home, and sometimes other extended family members such as a niece or grandson. The household was typically large (10 persons in 1860, 15 in 1870, and 11 in 1880), but this was not uncommon for the time and place.

Some censuses provide minimal information on the value of real estate and a household's personal estate. The McCoys' farm was valued in both the 1860 and 1870 censuses at $500. The household personal wealth was valued at $300 in 1860 and $200 in 1870. These valuations are not given in the 1880 census. When the McCoys' wealth was compared with that of a sample of 31 neighboring households in the 1860 census, the values for the Randall and Sally McCoy household was close to, but slightly below, the average of the sample.

The Pike County tax assessment rolls provided a glimpse into the farm's produce. Tobacco and corn were consistently important crops, varying from 4.5 to 68 kg (10 to 150 lb) of tobacco and 3.5 to 14 kl (100 to 400 bushels) of corn produced annually. Randolph McCoy was usually taxed for one horse and from one to 24 hogs, three to six cattle, and nine to 16 sheep annually. Trudy McCoy's (1976:146) account of the attack, abstracted from interviews with many of her descendants, suggests that the family had a full store of food (meat?) in the smokehouse at that time:

> The raiders were preparing to leave, when Tom Mitchell came out of the smokehouse where he had taken refuge after Ranel had shot his hand off. "Wait," he called. "The smokehouse is full of meat, molasses, honey, canned stuff, dried beans, apples, cushaws and pumpkins . . . enough to feed a regiment till spring. Burn the damned thing down."
>
> "Shore," one of the others agreed. "We didn't get to kill Ranel, let's burn up their foodstuff." Gathering a flaming board, he thrust it under the corner of the smokehouse. Waiting only long enough to see that it had caught, they [then] left the scene of their crimes. (McCoy 1976:146)

Mitchell's choice of words that the food would feed a regiment, versus a family, is also interesting, possibly reflecting his thinking of the attack as a quasi-military-style raid.

Information on the house is available from oral tradition only. Nothing remains on the site today except the stone-lined well. Most accounts of the attack on the McCoy house describe it as a one-and-a-half-story main house and a one-story detached kitchen connected by a covered breezeway. Both the kitchen and the house had a fireplace and loft (Dotson 2014; Jones 1948; McCoy 1976; Rice 1982; Spears 2015 [1888]; Waller 1988). The arrangement may have been a dogtrot form, as such structures typically have both pens of the same number of stories.

Trudy McCoy (1976:228) noted that family tradition held that Fanny McCoy said that during the attack she "stood in the doorway of the kitchen, not far from the well." Fanny McCoy's description of the event matches oral history from Jimmy Wolford, grandson of Randall McCoy's eldest son, Jim (McBride and Shipp 2015:13). During the site excavations, Wolford visited the site and recounted that his grandfather told him that the well was near the kitchen (this also seems a convenient and perhaps expected layout).

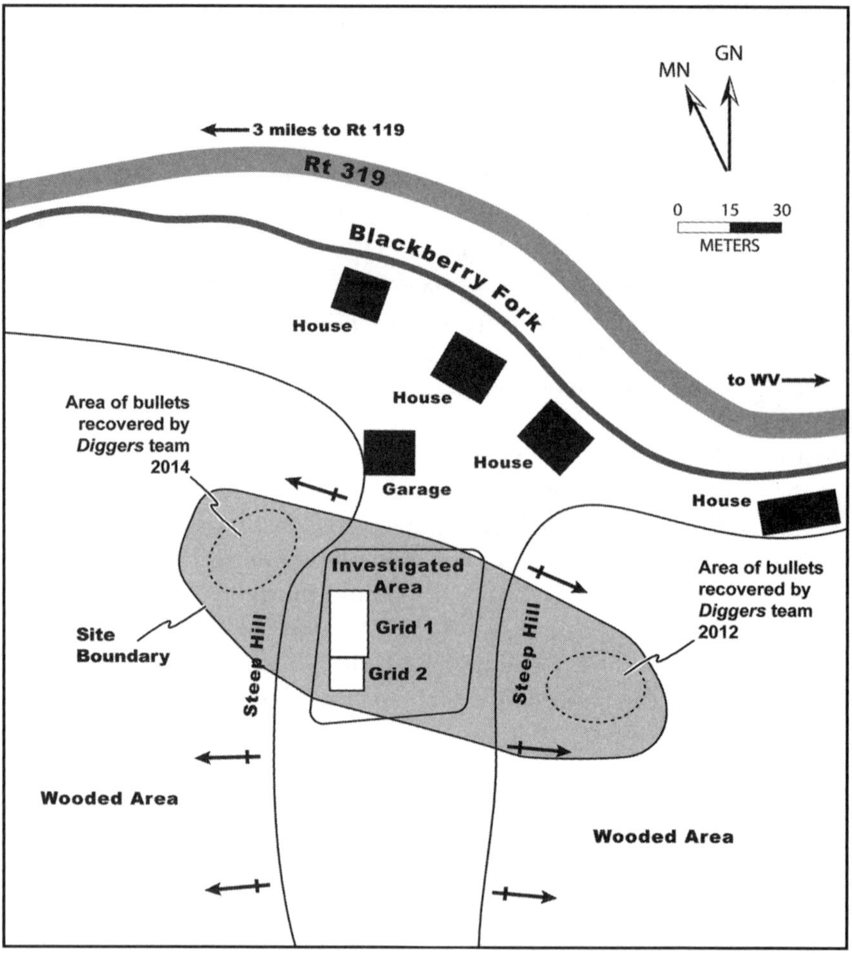

Figure 9.2. McCoy homestead site map. WV = West Virginia; MN = magnetic north; GN = grid north. (Courtesy of Kim A. McBride)

One final clue about the house location comes from a statement attributed to Sally McCoy, who is said to have wanted the graves of her three sons who were killed in 1882 to be visible from her front porch (McBride and Shipp 2015:52). These graves lie northwest of the site and are on the north side of Blackberry Fork, across today's Route 119. The graves can be just seen from the northern edge of where the house is believed to have been situated (Figures 9.2 and 9.3). John Spears's account of the attack, which he took mostly from testimony of the McCoys, refers to west (main house) and east (kitchen) portions of the house, as if the front was oriented to face Black-

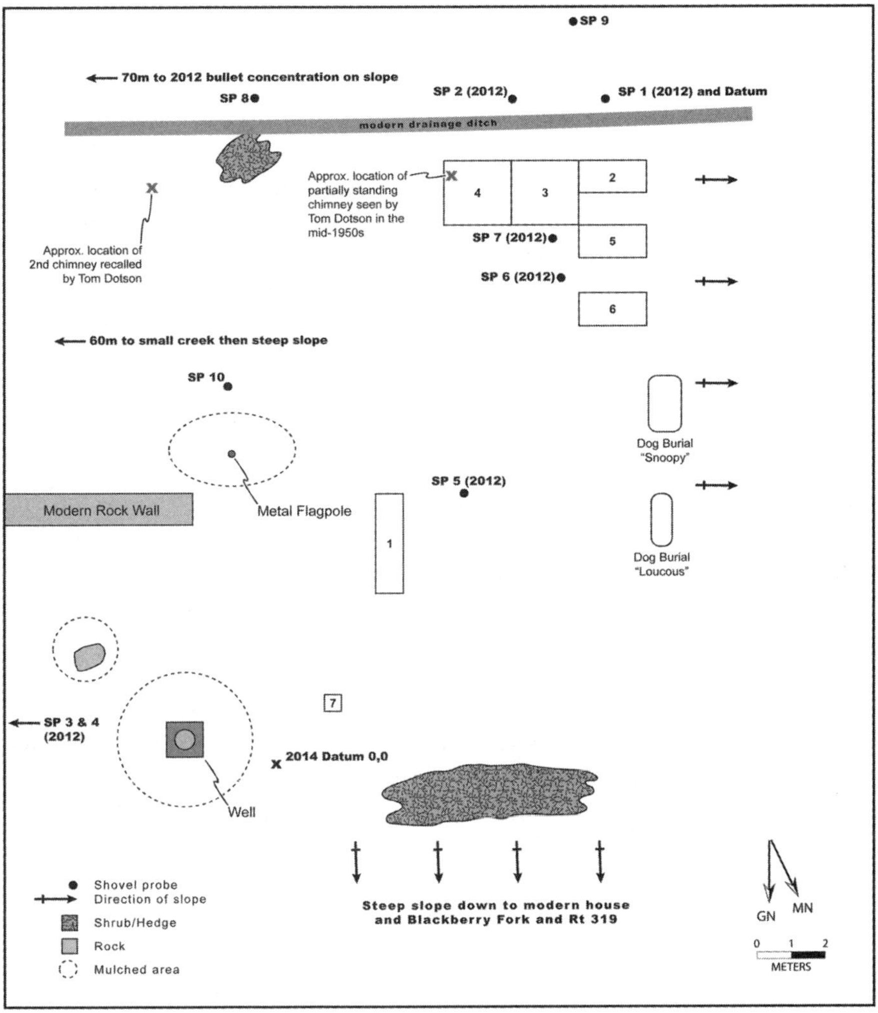

Figure 9.3. McCoy house, detail showing archaeological test units and shovel test probes (SP). 1 to 7 are test units. GN = grid north; MN = magnetic north. (Courtesy of Kim A. McBride)

berry Fork, which would have been the main local transportation artery (see Figure 9.2) at the time. This seems to be a logical and common arrangement.

The most compelling information about the house comes from Pricy Farley Scott. Pricy had grown up nearby and had visited the McCoy house before the attack. She later acquired the property. Pricy recalled that the McCoys' main house had four rooms with a loft overhead, and the kitchen had two rooms with a loft overhead (Dotson 2014:166, 245).

The Excavations and Battle Interpretation

The 2012 and 2014 archaeological investigations at the Randall and Sally McCoy site were conducted in conjunction with metal detecting carried out by the National Geographic *Diggers* television show stars "KG" and "Ringy." The *Diggers* team first visited the site for a few days in fall 2012 to search for Hatfield-McCoy feud artifacts to feature on their show. They discovered pieces of burned wood board, numerous nails, and miscellaneous pieces of metal, such as a probable plow blade and a possible cooking grate. All were found near the well that local tradition held was built and used by the McCoys (see Figure 9.3). The burned board was subsequently identified as southern yellow pine, a wood commonly used in frame house construction.

During their visit, KG and Ringy had also recorded four bullets in a concentrated area on the hillside southeast of the reported house location (see Figure 9.2). These bullets may be evidence of defensive fire from the McCoys inside the house toward the Hatfield attackers. Three of the bullets are .36 caliber and the other has a diameter of 0.3 inches; all had been fired. The Hatfield and Spence account mentioned Calvin McCoy firing a .38 caliber Winchester rifle and Randall firing a double-barreled shotgun loaded with birdshot. Although the feud accounts do not specifically describe anyone standing on the hillsides on either side of the house and kitchen, most of the accounts deal with the movements and actions of just a few of the attackers, primarily Jim Vance, Tom Mitchell, Cap Hatfield, and Johnse Hatfield. Certainly, other raiders may have moved around to take various positions encircling the house. If so, it would not be surprising if they had been shot at by Calvin and Randall McCoy.

The 2012 archaeological project was conducted several weeks following the *Diggers* team visit. Our goals were to confirm the site as the Randall and Sally McCoy farm, better define site limits and provide a basic idea of the nature of the deposits, and record the site (15Pi228) with the Kentucky Office of State Archaeology. We excavated 10 shovel test pits, several of which were expanded to about 50 × 74 cm (19.75 × 29.14 in), and conducted limited metal detecting.

Shovel tests excavated by the 2012 field team revealed both machine-cut and wire nails. We also reexamined the piece of slightly burned wood with an embedded nail that had been recovered by the *Diggers* team and identified the nail as hand wrought. Both hand-forged and machine-cut nails, although representing different periods of nail production, could have been in common use in this remote area in the 1850s. Four window glass fragments were recovered, two thin fragments (1.38 and 1.4 mm thick [0.054 and 0.055 in], respectively), and two thicker fragments (2.34 and 2.37 mm thick,

[0.092 and 0.093 in] respectively), perhaps suggesting at least two building episodes on the site. Recovered ceramics common to mid to late nineteenth century consisted of six whiteware and/or ironstone sherds and one sherd of salt-glazed stoneware. A large piece of dressed sandstone in the shape of an L, with evidence of shaping or dressing on both the inner and outer sides, could have been part of a chimney hearth firebox, or perhaps part of a pier support to the house. The recovery of a large piece of charred board, like that found by the *Diggers* team, suggested that archaeological remains at the site could include structural elements of the burned McCoy house and kitchen.

The soil stratigraphy across the site area suggested a complex occupational site history; observed strata including a mixture of "fill" clay soils containing ash and burned wood deposits. Conversations with present owner, Bob Scott, revealed that his father, Bankie Scott, had torn down a 1950s rental house in the 1980s that had been built on the property by former owners Crit and Pricy Scott. A bulldozer was used to smooth out and landscape the area (McBride and Shipp 2015:52). Despite this disturbance, the site appeared to have the potential to provide more information about the McCoy house and perhaps about the attack. Based on that finding, an archaeology field team returned in 2014, before and during a revisit by the *Diggers* team. A new goal of the 2014 investigations was to include Hatfield-McCoy descendants as visitors during the project. Over 100 descendants came from as far away as Oklahoma, New York, Florida, or the Carolinas to visit and participate in the excavations.

To gain more information on the location and orientation of the McCoy house and the pattern of the attack, ground-penetrating radar (GPR) was used to collect subsurface data that might give us a better idea of the house's location and footprint. Based on these findings excavation units were placed to examine the GPR-revealed anomalies. No intact structural remains were found, either from the GPR or from the excavations. In all, seven test units totaling 17.25 sq m (1,857 sq ft) were excavated (see Figure 9.3). Units 2 through 6 produced a complex matrix of burned sandstone remains, burned soils, and dense charcoal from the burned house and kitchen (Figure 9.4). Soil coring showed that this dark burned soil extended along the hillside to the north for about another 6 m (19.7 ft) past unit 6 (see Figure 9.3).

This area of structural debris yielded most of the artifacts recovered in the excavation and all the burned wood fragments. Assuming that the burned sandstone represents chimney rubble, it is likely that a chimney stood in the area of unit 4 where the sandstone was most dense and the soil exhibited the most intense evidence of burning. The location of this unit corresponded to an area where local resident and author Tom E. Dotson recalled seeing a chimney remnant in the mid-1950s (see Figure 9.3).

Figure 9.4. McCoy house sandstone chimney rubble. (Courtesy of Kim A. McBride)

Artifacts attributed to the McCoy occupation include 120 machine-cut nails; eight whiteware sherds and three porcelain and three stoneware ceramics sherds; the neck of a small patent medicine bottle; a mule shoe; a pulley; and a metal furniture pull. Ammunition recovered within the area of burned house debris consisted of three .22 caliber rimfire bullet casings. Of the three bullet casings, one is a rimfire short cartridge that is 1.07 cm (0.423 in) long. This cartridge type was developed in 1857 and is the oldest one still in production (Barnes 2000). The other two casings are from .22 Long rimfire cartridges. This cartridge type was developed in 1871 but was not widely available until 1887. One of these cartridges measured 1.5 cm (0.595 in) long and is still in production (Barnes 2000). The other measured 2.47 cm (0.975 in) long and is now obsolete. It was developed for the Winchester rimfire rifle or a Remington Special rifle (Barnes 2000). If Barnes's date of 1890 for this cartridge is correct, its presence postdates the 1888 attack. Both of the long casings had a stamped "U" on the head of the cartridge. This indicates that they were manufactured by the Union Metallic Cartridge Company, founded in 1867 in Bridgeport, Connecticut.

On their 2014 return to the site, the metal detecting team of KG and Ringy concentrated on the hillside to the west of the house. Seven lead bullets were recovered: three .30 caliber, three .25 caliber, and one too fragmentary for measuring (see Figure 9.2). Like the bullets found on the opposite hillside in 2012, these bullets most likely represent fire from the defending McCoys, and most likely Calvin with his .32 (from Mounts's account) or .38 caliber Winchester rife (from Coleman A. Hatfield's account), or Randall with his revolver (Spears's account).

No bullets were found in the immediate house area, which is surprising given the historical accounts of mass firing into the house by the Hatfield gang. This may be due to the disturbed nature of the site or to sampling bias as the excavated sample was very small. Future metal detecting by an experienced operator discriminating for lead (as there is a high density of iron objects at the site) might yield bullets from the attackers.

Thus, the specifics of the attack on the McCoy house played out like a typical raid in an irregular war fought by guerillas. Here a surprise attack in the middle of the night was launched by an armed Hatfield gang that greatly outnumbered the McCoy inhabitants. The Hatfields planned a quick raid, which turned into a brief but intense firefight and siege. The tactics used were those common to mountain guerrilla warfare of the American Civil War and thus were quite familiar to former Confederate guerilla leader Devil Anse Hatfield. In the raid, the Hatfield irregular tactics included using cover of darkness, catching the enemy by surprise, posting guards along the road, and then issuing a demand for Randall to come out of the house and surrender (possibly as a prisoner of war). Threats were then made to kill family members if he did not surrender or to set the house on fire to drive them out. The Hatfield gang then attacked weak points (doors), continued shooting, and used fire to the house and kitchen to force the family to surrender or attempt to escape. The bullets found on the slopes on opposite sides of the house locale suggest that they had placed men in good observation points to aid in the attack or shoot at anyone trying to escape.

The McCoys countered the raiders by barricading the door of the main house (where Randall was sleeping), although not the door of the adjacent semiattached kitchen, refusing to surrender (likely they did not trust the Hatfields), and returning gunfire. Calvin moved to the loft for improved observation and a better field of fire. He then escaped out the north side of the house where fewer Hatfields were positioned to create a diversion and allow his father, Randall, to escape. Finally, Randall fired, hitting both Cap Hatfield and Johnse Hatfield as he successfully escaped.

The outcome was the total destruction of the McCoy house, the killing of two and wounding of one McCoy family members, and the relocation of the McCoy family to Pikeville, away from the Hatfields' base in the Tug Fork Valley. The raid might be seen as an immediate tactical victory for the Hatfields. However, the aftermath for the Hatfields was strong and negative public opinion, loss of political support, and increasing efforts to capture them, all leading to their eventual relocation about 64 km (40 mi) to the east, away from the McCoys. In relocating, they sold their coal-rich lands on Grapevine Creek cheaply, just as the area was on the verge of major development and prosperity. Had they stayed in the area, they almost certainly would have sold these lands more lucratively or perhaps developed them within the family. One might state that the Hatfields won the battle of the McCoy farm, but in doing so lost the war.

The Raid in Historical Context

The attack on the McCoy homestead is much more understandable if one knows about the Civil War background of Devil Anse Hatfield and his men, especially their experience in Confederate guerilla raiding. Irregular or guerilla warfare was extremely common on both sides in eastern Kentucky during the Civil War and was, to some extent, officially supported by both Union and Confederate officers and their military policies (Mackey 2004; McKnight 2006, 2009; Mountcastle 2009; Noe 1997; Noe, ed. 1997; Noe and Wilson 1997; Sutherland 2009). However, until the 1864 repeal of the Partisan Ranger Act, which was intended as a stimulus for recruitment of irregulars into the Confederate States Army, the Confederacy provided more official support for partisan raiding than for independent guerrilla actions. But southeastern Kentucky had nothing like John Hunt Morgan from the Bluegrass of Kentucky or Mosby's Rangers of northern Virginia, who were Partisan Rangers with official backing from the Confederacy. Instead, the hostilities in eastern Kentucky were controlled by bands of Confederate-sympathizing guerillas or bushwhackers with little, if any, official affiliation or sanction.

Links between the irregular warfare of the Civil War and later eastern Kentucky feuds have been noted by scholars. For example, Brian D. McKnight (2006:233) suggests that "bushwhacking behavior might also hold the key to understanding the general lawlessness and the feuding tradition that would take hold in some isolated parts of the region in the decades following the war." William Lynwood Montell (1986:13), in a case study of killings in a Kentucky/Tennessee border area, notes that "by the mid-1870s the war-related violence had ceased, only to be replaced by general mayhem initiated and perhaps perpetuated in many instances by men who had learned the act of killing in guerrilla activity during and after the Civil War."

Both Randall McCoy and Devil Anse Hatfield had been members of John B. Floyd's Virginia State Line regiment, which operated like a partisan ranger unit, until it was disbanded on March 31, 1863. They subsequently both joined the 45th Virginia Infantry Battalion, commanded by Colonel Henry M. Beckley. Randall was captured on July 8, 1863, and was held in Union prison camps until June 16, 1865 (Hatfield and Spence 2003; Dotson 2014:109). Devil Anse's career was much different.

During his tenure in the 45th Virginia Infantry Battalion, Devil Anse was absent without leave by February 1864, as official records show (Dotson 2014). Accounts from Devil Anse's grandson Coleman A. Hatfield (Hatfield and Spence 2003) suggest that Anse deserted from this unit in late 1863 after refusing to follow orders to execute his uncle Slater Jim Hatfield and Slater's brother-in-law Anse Toler. Slater Jim Hatfield and Anse Toler had asked for leave to attend the deathbed of Toler's wife. When the leave was not granted, Devil Anse had advised them to go anyway, feeling that he could smooth over their brief absence. Instead, when they returned, Anse found himself ordered to execute them. Devil Anse then contacted another uncle (on his mother's side), James Vance, who was stationed nearby in another Confederate unit. Vance arrived with many horses and supplies and helped Devil Anse Hatfield, Slater Jim Hatfield, Anse Toler, and maybe others desert. Devil Anse may have also been motivated to leave official Confederate army duty and return home because the Unionist Home Guard had attacked his house in fall 1863 (Hatfield and Spence 2003:79–81). Because James Vance figures significantly in the Hatfield-McCoy feud, it is important to note that his Civil War career centered on raiding as part of John Wichter's cavalry and that he excelled in stealing horses (Dotson 2014; Hatfield and Spence 2003).

After his desertion from the army, Devil Anse Hatfield became the leader of an unauthorized guerrilla band. Devil Anse's own relatives describe him as "killing more people than consumption" during the Civil War. He was indicted for some Civil War deaths in Pike County, though he was never arraigned or prosecuted. Testimony of nineteenth-century West Virginia lawman Dan Cunningham suggested that Devil Anse led raids on at least 19 houses and killed at least six men during the Civil War. Devil Anse collaborated with a more famous local guerilla, William "Bill" Smith, especially on larger operations such as when attacking shipping or when assisting Confederate forces at the Battle of Scary Creek (Dotson 2014; Hatfield and Spence 2003; Johnson 1986).

As these guerilla raids often involved taking plunder, Devil Anse likely accumulated wealth during the Civil War that helped him develop his timbering operations after the war. In contrast, the data presented earlier concerning Randall and Sally McCoy's farm suggests nothing above a middle-class farm operation. Randall, as a prisoner of war, had no chance to benefit from the war financially, unlike Devil Anse.

Also relevant to the Hatfield-McCoy feud is that Devil Anse had many interactions during the Civil War with persons who would later play a part in his hostilities against the McCoys and their allies. For example, Devil Anse had several negative encounters during the war with Colonel Dils, a Unionist. Dils had been dismissed by President Lincoln from the 39th Kentucky Mounted Infantry in December 1863 for seeing too much to his own financial interests. His popular support at home, however, led to his forming a Unionist Home Guard made up in part of followers who had deserted from the 39th. In this capacity he and Devil Anse Hatfield were often at odds. Some of the wartime antagonism between them resulted from several occasions when Hatfield's men captured and burned Dils's goods or store buildings. Hatfield likely also participated in the Confederate guerilla raid that captured Dils and held him for ransom (Dotson 2014:92, 97).

Another Civil War event that possibly increased hostilities between the McCoys' allies and the Hatfields was the 1865 murder of Randall McCoy's brother Asa Harmon McCoy. This should not be seen as part of the "feud," however, which in any cohesive sense did not start until the August 1882 killing of Ellison Hatfield (Sr.) by Randall's sons Tolbert, Pharmer, and Bud McCoy. Although Randall McCoy was a Confederate soldier, his brother Asa Harmon and many of his neighbors were Union supporters. Asa Harmon had served under the command of William "Bill" Francis (sometimes France) of the Unionist Home Guard of Pike County. Hatfield family accounts suggest that Devil Anse wanted Asa Harmon dead, having heard that Asa Harmon had planned to kill him in retaliation for his killing of Bill Francis in the fall of 1863. Thus Devil Anse's killing of Asa Harmon McCoy might be seen as taking preemptive defensive action, much like the raid on the McCoy home. It is likely that the killing of Bill Francis was accomplished not by Devil Anse Hatfield himself but by members of his band, possibly including James Vance or Jim Wheeler Wilson (Hatfield and Spence 2003:85–86). Tom Dotson (2014, 2015a) suggests that Vance has been unfairly framed for this event and many others. He was killed near his home on Thacker Creek in Logan County on January 7, 1888, by a Kentucky posse led by Frank Phillips, a son-in-law of Asa Harmon McCoy.

After the Civil War, Devil Anse continued to be recognized as an area leader who could raise men to act on his or other's behalf. His oldest son (who participated in the raid) was called Cap, or Little Cap, in honor of Devil Anse being a military "captain," although officially, when in the Virginia State Line or 45th Battalion, Anse rose no higher in rank than a lieutenant. Captain was an honorific address often used locally in the nineteenth century for men considered adept in military service. After the war, Devil

Anse could still mobilize groups of men to act in a disciplined military-like force, as he did in 1886 when he brought a group of over 30 armed men to the Logan Courthouse to "subtly" influence residents to vote for the political candidate that he supported (Dotson 2014).

Randall McCoy's antagonism of Devil Anse Hatfield is sometimes portrayed as having been inflamed by his jealousy over Hatfield's economic success, achieved largely through timbering (Dotson 2014; Sparks et al. 2015; Waller 1988). Randall McCoy had also been briefly involved in timbering, but without any financial success (Waller 1988). As early as 1988, Altina L. Waller suggested that control of timber and coal resources was the real motivation of this stage of the feud. This explanation has been taken up more recently by Dotson (2014), who cites the timely influence of the 1887 news of the Norfolk and Western Railroad coming up the Tug Valley (see Figure 9.1) and the release of new geological survey maps showing large coal reserves on Devil Anse Hatfield's Grapevine Creek lands in West Virginia. At a time when "coal was king" or at least on its way to dominance, this geological information may have encouraged Perry Cline and John Dils in late 1887 to break a five-year interval of peace and step up efforts to capture the Hatfields for the 1882 killings of Tolbert, Pharmer, and Bud McCoy. These captures, in this explanation, could have been used to encourage Devil Anse Hatfield to sell the coal-rich lands he held on Grapevine Creek. As Anse did sell shortly after the McCoy farm attack, in this sense the raids into West Virginia achieved their objective. This sale was key to ending the feud (Dotson 2014; Hatfield and Spence 2003; Waller 1988).

As for the immediate motive to raid the McCoy farm, Dotson (2014) and Hatfield and Spence (2003) suggest that the prime objective of the January 1, 1888, raid was to capture, not kill, Randall McCoy, on the assumption that Randall could be used to force a trade for some of the Hatfields held by McCoy allies in Kentucky. The Spears history and Ellison Mounts's confession suggest this to be the case. The raid on the Randall and Sally McCoy farm came after several months of Kentucky-based incursions into Hatfield lands in West Virginia by bounty hunters and detectives, especially Frank Phillips and his men. Phillips was a known, local "tough guy" who happened to have been, like Perry Cline, a ward of Colonel Dils. Phillips had been appointed by Kentucky governor Buckner to apprehend the Hatfields for the 1882 killing of Tolbert, Pharmer, and Bud McCoy. Those Hatfields held in the Pikeville jail awaiting prosecution included Devil Anse's brother Wall. Many of the Hatfields, under threat of capture by "deputies" or bounty hunters, had been hiding out for months and had grown very weary of this state of affairs, a sentiment even expressed by Devil Anse (Hatfield and Spence 2003).

After the Battle

The raid on the Randall and Sally McCoy farm was one of the more dramatic and brutal events of the feud, especially with its unusual death and injury to women. The brutality of the raid caused it to receive quick coverage by West Virginia, Kentucky, and Ohio newspapers, and it eventually became a national story. This was extremely bad news for Devil Anse, as it eroded his support locally and within the West Virginia government. This loss of local support ultimately led him to relocate farther east to the area of Main Island Creek near Logan, West Virginia.

Before the 1882 Election Day killing of Ellison Hatfield (Sr.), hostilities between the Randall McCoy and Devil Anse Hatfield families had not been seen as a feud and were not the subject of widespread media coverage. After the raid on the McCoy farm, however, reporters such as T. C. Crawford of the *New York World* (Crawford 1889) visited the area and wrote extensively about the feud and feudists, beginning the process that would ultimately make the Hatfield-McCoy feud known around the world. Some scholars have recently portrayed the sensationalistic newspaper and affiliated "dime novel" coverage of the January 1, 1888, attack as central to the late nineteenth-century development of stereotypically negative portrayals of Appalachian residents as backward, ignorant, and isolated, and even as the start of "yellow journalism" (Bonfig 2016; also see Hanmer 1997).

Conclusion

The Hatfield-McCoy feud seems to remain relevant today in many arenas, as for example its recent use by scholars wishing to study destructive leadership styles (Sparks et. al. 2015). In this context, both Randall McCoy and Devil Anse Hatfield are identified as examples of destructive leaders: such leaders are seen as often exhibiting charisma, personalized power, narcissism, negative life themes, and ideology of hate, especially in environments of instability, perceived threat, lack of checks and balances on power, and ineffective government institutions. This environment fits the setting of the Tug Fork in the 1880s. George Sparks, Patricia Wolf, and Andryce M. Zurick, writing in 2015, see Randall McCoy as an especially destructive leader for his role in the killing of Bill Staton (brother-in-law to Ellison Hatfield) by two of his nephews and for his repeated legal actions against the Hatfields. In turn, Devil Anse Hatfield is seen as an especially destructive leader because of the events of the killing of Tolbert, Pharmer, and Bud McCoy in 1882 and the January 1, 1888, attack on the McCoy farm.

Otterbein (2000; following John Ed Pearce 1994) argues in his study of

Kentucky feuds that their ultimate cause was economic. Neither honor, nor revenge, often popularly associated with the cause of feuds, were the primary reasons for feuding among the Hatfields and McCoys. In this case he suggests that control of illegal whiskey may have been an underlying trigger. Waller (1988) and Dodson (2014) appear to provide evidence for Otterbein's thesis, but in their interpretations of the Hatfield-McCoy case, it was competition for control of extremely valuable timber and coal resources. A similar argument has been made for another famous Kentucky feud, the Baker-Howard feud of Clay County, but in this case the resource in question was salt (Blee and Billings 1996; Billings and Blee 2000).

As it turns out, the raid on the McCoy farm was located in the same region of the country where Native Americans and colonists raided and ambushed each other throughout the eighteenth century (see McBride and McBride, chapter 1). The same tactics were also used. This brings to the fore the interplay between culture, environment, and circumstance in low-intensity conflicts. This is a topic for another study of the iconic Hatfield and McCoy feud.

Acknowledgments

Excavations were undertaken by the Kentucky Archaeological Survey, with funding from the National Geographic Foundation, Half Yard Productions, City of Pikeville, and Pike County Tourism, in association with the National Geographic television show *Diggers*. Thanks to landowners Bob and Rita Scott, Pike County Tourism, and the many Hatfield and McCoy descendants for help on-site.

10
A Ukrainian Insurgent Army Company Ambush of a Soviet NKVD Battalion, 1945

A Multidisciplinary Study

Adrian Mandzy

Of all the territories under Soviet and Nazi occupation in the twentieth century, it was in Ukraine that nationalists mounted one of the largest and longest lasting asymmetric movements in their efforts to seek independence. Scholars estimate that the Ukrains'ka Povstancha Armia (UPA), often referred to as either the Ukrainian Insurgent Army or the Ukrainian Partisan Army, fielded approximately 28,000 combat troops. Recent research suggests that as many as 2,500,000 individuals may have provided support to the UPA (Patryliak 2012). Significantly, for most of its existence the UPA was not supplied by any foreign power; the overwhelming majority of its arms and supplies were captured from the enemy. Despite this fact, the offensive operation launched by the UPA troops in January 1945 demonstrates the combat effectiveness and the outstanding logistical support system of the insurgents (Figure 10.1).

From the 1920s until the early 1960s, individuals dedicated to the cause of an independent Ukrainian state waged military operations against Polish, German, and Soviet authorities. The longevity of the conflict and the widespread involvement of the local population against these outside forces provide a powerful example that contributes to our understanding of the dynamics of asymmetric warfare. Innumerable military actions were undertaken by the UPA, and Myroslav Symchych's ambush of a Soviet battalion in 1945 is one instance of an offensive operation launched by a UPA company against a much larger force.

The UPA evolved from earlier efforts at national liberation. Following the failed attempt to create an independent Ukrainian state in the aftermath of the collapsed Russian and Austro-Hungarian Empires, Ukrainians found themselves living under various political regimes. The majority of ethnic Ukrainians were incorporated into the Soviet Union, whereas others were

Figure 10.1. Map of Ukraine in Europe. (Courtesy of Volodymyr Chornovus)

ruled by Poland, Czechoslovakia, Hungary, and Romania. Uncompromising in their devotion to the ideal of a Ukrainian national state, a small handful of Ukrainian veterans continued to push for political independence (Kosyk 1993:14). Recognizing that creating and supplying a traditional military force of battalions, regiments, divisions, and corps was extremely difficult when facing a much larger, established, conventional army, these revolutionaries sought other methods to achieve their goal of an independent Ukraine.

Following World War I, the Polish government, eager to reduce the Ukrainian presence on its eastern frontiers, used a variety of tactics, including restrictions on language and education, forced religious conversion, and distribution of state lands to Polish settlers, to create a monolithic Polish state (Borovych 1986:62). Throughout the 1920s, the Ukrainian Military Organization (UVO) employed terrorist tactics modeled after those successfully pioneered by Michael Collins in attaining Ireland's independence from Britain. The Polish regime responded by launching a pacification campaign against various established Ukrainian institutions (Wallace and Chyz 1931). The conduct of the pacification campaign resulted in more Ukrainians supporting and joining the radical nationalists. Taking advantage of its notoriety among the new postwar generation, the UVO sought to merge the young Ukrainian revolutionaries into a new entity, the Organization of Ukrainian Nationalists (OUN). By the early 1930s the OUN had replaced the UVO, and its members committed various terrorist acts—assassinations, arson, and robberies—against both the Polish and the Soviet governments from which they sought independence (Posivnych 2010).

As early as 1934, tensions in the OUN began to surface between older members living abroad and those younger members living within Ukraine. With the assassination of Ievhen Konovalets, the leader of the OUN, by a Soviet agent in 1938, the internal friction grew, and the younger generation of resident OUN members began to take on greater responsibilities. In early 1939, OUN volunteers crossed into Czechoslovakia and fought for the short-lived Transcarpathia. Imprisoned by the invading Hungarians, several OUN members were turned over to Polish authorities who either executed them on the spot (Onyshchuk 2013) or sent them to Polish prisons.

In September 1939, Nazi Germany and the Soviet Union attacked Poland. The Soviet advance into western Ukraine, although touted as "Ukrainian liberation from Polish rule," actually brought little relief for the local inhabitants. The Soviets initially dismantled the existing Polish political, socioeconomic, and cultural infrastructure, which had the greatest impact on the middle-of-the-road liberal organizations. By spring 1940, mass arrests and deportations of Poles, Jews, and Ukrainians by Soviet authorities resulted in thousands of alleged "enemies of the people" being packed into cattle cars and sent to work in slave labor camps in Siberia and Kazakhstan (Subtelny 1991:456).

In the wake of the German and Soviet invasion of Poland, many OUN members escaped from prison and renewed their struggle for national liberation. By 1940, the OUN had split into two factions—an older generation dominated by veterans of 1917–1920 (OUN-M), and the younger radicals who joined the organization in the 1930s (OUN-B). In an effort to find needed support for their resistance to Soviet persecution, both factions of the OUN sought to increase cooperation with the Nazi leadership. Two independent OUN units, Nachtigal and Roland, were formed for the upcoming "Crusade against Communism" (Bol'novs'kyj 2003).

When Germany invaded the USSR in 1941, the retreating Soviets executed as many "undesirables" as they could. The Nazis publicized these executions and began to conduct their own mass executions of civilians. In the ensuing chaos, on June 30, 1941, the younger generation of the OUN proclaimed an independent Ukraine (Trident 1941). The German authorities were angered and responded by detaining and shooting those OUN members they could locate (Ilnytzkyj 1956:338–339). Those "fortunate" enough to be arrested were sent to concentration camps.

Armed resistance to Soviet authority began sporadically with attacks against retreating Red Army troops, political prisons, and mobilization centers. In 1942, attacks on centers of the Nazi regime occurred haphazardly, and armed resistance increased in the Volyn' Region and the swamplands of the Polissia Region. It was in this area that Taras "Bul'ba" Borovets organized a group of Ukrainians that fought against both Soviet partisans and German troops (Subtelny 1991:473). In December 1941, Borovets adopted the name Ukrains'ka Povstancha Armia (UPA) for his group.

Initially, both factions of the OUN were ill prepared to fight against the large, well-supplied Soviet and German armies. By early 1943, however, self-defense companies (OUNSD) and OUN-M units began actively eliminating German outposts and battling marauding Soviet partisans. By July 1943, there was a recognized need to unite the three Ukrainian factions (OUN-M, OUN-B, and OUNSD), but all attempts at dialogue failed. Over the course of the summer, OUNSD units surrounded the other partisan groups and disarmed them. Many of the disarmed units were ultimately integrated into a new structure, the unified UPA. By the end of 1943, the UPA had effectively absorbed all other independent Ukrainian groups.

In 1944, in the mountainous regions of the Carpathians, OUN members began to form units of the UPA. In fall of that year, a newly trained UPA officer, Myroslav Symchych, was assigned to the Birch Company in command of a platoon. By the beginning of 1945, the 22-year-old comman-

Figure 10.2. Myroslav Symchych in 2011 at the site of the 1945 ambush. (Courtesy of Adrian Mandzy)

der Symchych was second in command of the company and served as the unit commander during the 1945 ambush of a Soviet battalion (Figure 10.2) (Myroslav Symchych, personal communication 2011).

Within the Carpathian Mountains, the village of Kosmach and its surrounding hamlets had become a supply and administrative center for the UPA. In January 1945, a Soviet security force was sent in to eliminate the center of the insurgents. During the ensuing engagement, the NKVD forces wired for reinforcements, and an additional unit was sent to help with the raid.

According to the now almost 90-year-old Symchych, he set up his ambush a few kilometers outside of town. Symchych disbursed his troops along the two ridgelines that overlooked the road at a natural choke point—a mountain stream (Myroslav Symchych, personal communication 2011, 2012). Arriving on site before the enemy, Symchych's men dismantled the small wooden bridge that crossed the stream. When a small group of Soviet scouts approached the stream, the insurgents held their fire and were not spotted

Figure 10.3. Terrain at ambush site. (Courtesy of Adrian Mandzy)

by the NKVD troops. As the Soviet scouts crossed the stream and went on ahead, Symchych sent a group of his men to follow them and keep the NKVD troops from returning and turning his flanks. Sometime later, when 12 heavy troop trucks and one automobile belonging to the NKVD came to the missing bridge, the vehicles stopped. The troops dismounted, formed into two columns, and prepared to advance on foot to the village. At this point Symchych gave the command to fire, and his company destroyed the battalion of Soviet soldiers and killed the commanding officer. Later it was discovered that the commander was Major General Nikolaj Dergachov, best known for his punitive actions against Chechen and Tatar civilians. Wounded in the arm during the action, Symchych later collapsed from the loss of blood and was never able to give the command to come down from the hills and collect weapons from the destroyed convoy (Figure 10.3).

Imprisoned in 1949, released in 1963, imprisoned a second time in 1968, and finally released in 1985, Symchych has told his story many times and has become a local celebrity. A small monument was erected on the spot where he commanded the troops in the ambush, and two television documentaries have been made about his story. A second (3 m tall) monument commemorating the action was raised in the neighboring town of Kolomyia.

Yet the story of the ambush is not common knowledge, even among scholars studying the UPA (Tys-Krokhmaliuk 1972). The hundred plus volumes of the *Litopys of the UPA* (Chronicle of the UPA) are also rather silent on the engagement. Soviet sources mention the death of the general but downplay its significance. Accounts given by the 90-year-old Symchych, almost 70 years after the event, may also be inaccurate, the details being lost in the fog of time.

The Archaeological Evidence

The developing methods of battlefield and conflict archaeology, when applied to circumstances such as those in the Kosmach ambush, can provide additional sources of information that validate, fill in, and clarify the vagaries of memory. Mid-twentieth-century firearms of the sorts attributed to the 1945 action expended many metal cartridges when fired. Made of brass, copper, steel, or copper-cased steel, these artifacts, when recovered using proper metal detecting and other archaeological methods of spatial control, can provide a significant amount of data on a military action that may be absent from the historical record. The cartridges make it possible to determine not only the caliber of a weapon in play on the battlefield but also, often, the date and place of manufacture, through information stamped onto the heel of the cartridge (Hogg 1982). In addition, over the last 30 years, field studies have demonstrated that documenting the disbursal of cartridges across the terrain of a battlefield can provide solid evidence for how individuals moved and behaved. Imprints from extractors and firing pins can also be compared, and it is possible to determine whether two similar cartridges were fired by the same or different individual weapons (Scott and Haag 2009). The presence of deeply indented primers can suggest that the weapon fired from an open bolt, such as a fully automatic machine gun. The advantage of an open bolt design is that it removes a separate hammer/striker and spring and allows for airflow into the weapon's chamber, which cools the firearm.

Events that have impacted a battlefield landscape since the time of the action must be taken into consideration when conducting field studies and analyzing the findings. In a mountain environment, erosion from spring runoffs can significantly change a landscape. As the Kosmach ambush took place in the middle of winter, the Ukrainian troops in the ridges above the ambush site were not dug in, and any expended ordinance would have dropped into the snow. Given the nature of the battlefield terrain, as the spring thaws melted the snow, many of these cartridges could have been moved or simply washed away. When compounded by 70 years of such activities, the damage to certain of the archaeological deposits created by the action can become significant.

In the valley area held by the NKVD, the construction of an earthen mound has shifted or covered physical evidence of the encounter. In addition, the construction of a new road at the bottom of the hill and its periodic springtime flooding would also have a negative impact. During an initial visit to the battlefield in 2011, two distinct rebuilt road sections and three bridge construction sites were noted by the author.

In 2012, a field team composed of students and volunteers conducted a two-day survey of the ambush site (Mandzy 2007, 2012). The team included such notables as the former leader of the 1945 ambush, acting company commander Myroslav Symchych; local government officials; the director of the historical museum Prison on Lontskoho, Ruslan Zabilyj; and Jaroslav Onyshchyk of L'viv University. This small group recovered more than 200 artifacts. As each artifact was recovered, it was given a unique number, and its coordinates were taken by using a handheld GPS instrument. These coordinates were later transferred to a map in ArcView to prepare distribution maps key to interpreting military positions occupied during the ambush.

Given the metal detecting strategy used in the field study, most of the recovered items were spent cartridges, some steel ammunition clips, and miscellaneous iron objects. Much like the 1980s battlefield survey conducted by Douglas D. Scott and Richard A. Fox Jr. at the Little Big Horn National Battlefield, our objective was to identify firing positions established by the insurgents, and, if possible, to trace the movement of troops on the battlefield over the course of the action (Scott and Fox 1987). According to Symchych, he ordered his men to stay in place once set up and not to move. In an interview given in 2002, Symchych stated that his forces had 22 light machine guns, a regimental mortar, 20 automatic weapons, 15 semiautomatic rifles, and an assortment of bolt-action firearms. In a 2011 interview, he stated that his troops were also armed with a 20 mm antitank weapon (Ovsienko 2002; Myroslav Symchych, personal communication 2011, 2012).

Our study divided the 1945 battlefield area into three zones: the eastern slope, the western slope, and the central valley ambush area. The central valley ambush area was significantly altered over the last 60 years principally by the construction of a new road and the building and rebuilding of new bridges damaged by spring floods. At least three different new bridge building phases were identified in the field.

Our survey of the ambush area uncovered various pieces of unidentifiable iron scattered across it. Although some of these items may be parts of the vehicles targeted by the ambush, the only definitive military items recovered were a small twisted fragment of lead, which perhaps splintered off of a larger bullet, and a clip of five unfired Austrian rounds from World War I. It is extremely unlikely that the Soviet NKVD troops were using World

War I surplus ammunition in 1945, and its recovery in the ambush kill zone makes it unlikely to have been dropped by the UPA insurgents during this particular engagement. More than likely, the clip was lost sometime during the previous world war.

Our survey recovered 121 artifacts across the east ridge study area. The recovery of 102 cartridges in a 0.5 m (1.6 sq ft) square area suggested a machine gun emplacement. Symchych stated that his company had light machine guns, and the material evidence confirmed that claim. The 102 cartridges were all 8 × 56R caliber and marked *AH* or *ML*. The initials *AH* are the mark of the Csepel Arsenal company in Budapest, Hungary, and the initials *ML* refer to the Magyar Loszermuvek RT in Veszprem, Hungary (Nowak 2003:47). Both companies worked to supply the Hungarian military between World War I and II.

Like most military ammunition, these Hungarian-manufactured rounds were marked with their year of production. Those found were made in the following years: 1934, 1936, 1938, 1939, 1940, and 1941 (Table 10.1). The arsenal markings and dates suggest that this ammunition came from a variety of sources and not from just one crate of ammunition. In autumn 1943, when units of the Hungarian army were assigned to guard railroads and highways for the Germans, they established a secret armistice with the UPA. This armistice "provided the two sides with assurance of mutual non-aggression, exchange of information, together with foodstuffs in return for arms and ammunition" (Sodol 1987:21). These cartridges were most likely a product of that agreement.

The indentations made on these Hungarian-made cartridges were identical, which demonstrates that they were all fired by the same weapon. The deep penetration of these strikes on the primer indicate that they were fired by a weapon with an open-bolt mechanism. The quantity of the rounds recovered from one very small area when considered along with the previous cartridge analyses indicates the discovery of a machine gun position, but which particular weapon was used? Several heavy and light machine guns were chambered for the 8 × 56R cartridge. The Schwartzlose heavy machine gun, developed for the Austro-Hungarian Army before World War I, originally used the 8 × 50 cartridge. The later model 07/31, however, used the 8 × 56 cartridge (Hungarian Ring ca. 2015); that cartridge was also used in two light machine guns—the Bulgarian ZB-39, which was a refined version of the Czech Bren gun, and the Hungarian Solothurn Model 31 machine gun.

The recovery of Hungarian-produced ammunition discredits the ZB-39 as a source. Bulgarian troops were predominantly involved in the Balkans, and it is unlikely that one of the 500 or so manufactured ZB-39s made it into

Table 10.1. Spent Hungarian Cartridges Recovered from the Machine Gun Position

Year	AH mark	ML mark	Total
1934	5	0	5
1935	0	0	0
1936	6	0	6
1937	0	0	0
1938	8	0	8
1939	35	0	35
1940	11	27	38
1941	0	10	10
Total	65	37	102

Note: AH, mark of the Csepel Arsenal company; ML, mark of the Magyar Loszermuvek RT.

the hands of the Ukrainian insurgents operating in the Carpathians. Similarly, the belt-fed Schwartzlose heavy machine gun weighed more than 22 kg (48 lb), which made it poorly suited for mountain warfare. The Solothurn Model 31, however, was commonly used by Hungarian troops in Ukraine. The single known photograph of Symchych's company shows several such weapons (V'iatrovych et al. 2011), and it is most likely that the UPA insurgents used a Solothurn Model 31 at this location.

Additional 8 × 56 caliber rounds were found along the eastern slope study area, but their firing pin imprints were different from those on the cartridges attributed to the machine gun position. The strike imprints on these cartridges were shallow and suggest the use of a bolt-action rifle. Differences in the sets indicate fire from two riflemen who were most likely using Hungarian Model 31 Mannlicher Steyr carbines.

The recovery of 7.62 × 54R cartridges indicates the presence of Imperial Russian or, more likely, Soviet weapons. The Mosin rifle, also known as the "three-line" or the Model 91, fired this cartridge. Entered into production before the Russo-Japanese War, it was produced in the millions. Large numbers were lost in World War I. A slightly shorter model of the rifle, the 91/30, was also produced in record numbers, and many of these were lost in the early days of the Nazi advance into Poland and Ukraine. The Soviet carbine versions of the Mosin—the M38 and the M44—and the Soviet semiautomatic rifles AVT 36, SVT 38, and SVT 40 also used the 7.62 × 54R cartridge.

The recovered 7.62 × 54R cartridges significantly predated the beginning of World War II. These cartridges dated to 1938, 1939, and 1940, with only one cartridge dated to 1944. The varieties of metals used in the manufacture of Soviet cartridges resulted in poor preservation, and most of those found were lacking clear head-stamp information.

One 7.62 × 54R caliber cartridge was marked with the letters "Sh." This cartridge was a high-quality round originally developed for the ShKAS aircraft machine gun. As the ShKAS became obsolete, snipers in the Red Army used the ammunition for its uniform velocity and superior primers. The round carried a special explosive ZR bullet, which marked the target with a puff of white smoke, making it desirable to snipers.

Test firings of an SVT 38 and an SVT 40, as well as various Mosin model weapons, revealed no evidence of a specific firing pin depth for the individual weapon types. As the bolt of any Mosin can be adjusted to increase or decrease how far the firing pin protrudes, the measurement of firing pin strike depth does not allow reliable identification of the Soviet, or even tsarist, weapon or weapons used in the ambush. Although it is impossible to state which Soviet weapons were used, analysis of the firing pin imprints does indicate the presence of three more riflemen in action on the eastern ridge, one of whom may have been using sniper ammunition.

The recovery of three 7.62 × 25 cartridges, with identical firing pin impacts, indicates the use of a Soviet submachine gun or possibly a Tokarev TT pistol. Each of the three cartridges was different; one was marked with a "T" and "44," a second with a "3" and "44," and the third was unmarked. Although it has been suggested that unmarked 7.62 × 25 cartridges were produced in relatively small batches for the Tokarev TT pistol before 1939, such cartridges were used by the Soviets in large numbers in the later stages of World War II.

In the western slope study area, cartridges found identified the positions held by six more insurgents. One rifleman used a Mannlicher rifle and expended a minimum of 38 rounds. In addition to these fired cartridges, the insurgent left behind three metal clips from his weapon. In a Mannlicher Model 95 Streyr rifle or carbine, the clip containing the ammunition is inserted into the weapon. Once the fifth round is chambered, the clip drops from the bottom of the weapon and can be easily lost even during peacetime at a firing range. In the chaos of combat and in very cold conditions, it was inevitable that soldiers would lose these clips (Figure 10.4).

Among the rocks in the western study area, two additional metal pieces were recovered. The first is oval and is approximately 3 mm (.18 in) in width. Although such ovals were used in a variety of items, the dimensions are the same as the lower sling of a Mannlicher rifle. The second object is in the shape of a large Latin letter U and is similar to the iron wire hangers found on some Mosin rifles. Similar U-shaped attachment rings were also used on

Figure 10.4. Rifleman's firing position. (Courtesy of Adrian Mandzy)

German assault harnesses, and this item may have been part of a combat harness. The lack of any post-1945 military artifacts on either the east or west ridges is a good indication that these iron pieces relate to the ambush, and the proximity of the two items suggests that they may have both been part of the same harness or strap.

A second soldier on the western slope was armed with an 8 mm (7.92 mm) weapon. These steel-cased cartridges were used by both Germany and Hungary during World War II, and it is most likely that this soldier was armed with some type of Mauser. The German 98k was the most common and could easily have been captured during an action with Nazi troops. Other types of 8 mm weapons were also in service, such as the 8 mm Hungarian Model 43, although it was the less common of the two.

Other cartridges recovered identify two other soldiers armed with light Soviet submachine guns. These were most likely either the PPSh 41 or PPS 43. Three other troopers were armed with 7.62 × 54R caliber ammunition, which could have been fired by a Soviet self-loading weapon or any Mosin variant.

Documenting the location of each individual find with GPS allowed us to better understand how the artifacts relate to one another and the individuals using them. Using a traditional transit to record the location of the artifacts in this mountainous terrain would have been difficult, unreliable, and dangerous. We plotted the location of each find within 3–5 m (9–16 ft) of its recovery. The spatial patterning of the recovered artifacts revealed how the January 1945 ambush was set up. All artifacts were recovered from the middle of the ridgeline study areas where the slope was the least steep.

The largest concentrations of spent ammunition was collected in niches between rocks. In areas of steeper slope, fewer finds were recovered. A contemporary photograph of the western ridge illustrates the terrain topography of the ambush site and why no artifacts were recovered from this slope. At this exposed location, the gravel continues to erode out of the mountain, and no vegetation is able to put down roots. Any cartridges deposited in this location would not have been preserved.

The Documentary Evidence

At first glance, given the lack of documentary evidence, interpreting the operations of the Birch Company of UPA partisans in the 1945 Soviet ambush appears to be rather difficult. However, it is possible to reconstruct their conduct in the field using photographic evidence, contemporary UPA sources, and oral accounts from the field commander.

While evolving from a small local organization to a large-scale national liberation movement, UPA needed to rapidly increase the military training and effectiveness of newly raised combat troops. In various parts of the country, schools for NCOs and junior officers were set up and taught by senior UPA members who had either gained combat experience in the aftermath of World War I or who had served within the ranks of the Polish, German, or Soviet forces. To provide these new leaders with the tools of the trade, the UPA High Command published manuals on how to conduct field operations. Four separate hard-covered volumes appeared under the title *Boyovyi pravylnyk pikhoty* (Infantry combat manual). Published in 1944, the UPA manuals are a translation of a 1943 Red Army manual published in Moscow under the same name (Sokhan, Shtendera, and Ivanenko 1995). The first volume deals with the obligations and responsibilities of the individual soldier, whereas the other three volumes provide rules for the operation of squads, platoons, and companies.

Although the military doctrine noted in the UPA manuals is rather broad in scope, it does provide a point of reference for interpreting the ambush at Kosmach. For example, according to the soldier's manual, "Heavy machine guns and antitank weapons are to be used to provide fire on the flanks and between platoons" (Sokhan, Shtendera, and Ivanenko 1995:320). The concentrated recovery of a quantity of expended machine gun cartridges appears to define the placement of a machine gun on the northern flank of the eastern ridge.

The manuals address different types of infantry units and their proper actions for offense and defense. Discussed are units employing rifles, submachine guns, machine guns, antitank rifles, and individual artillery pieces. Not all the information follows the same format. For example, the rifle squad manual focuses on how to deploy from a single-file marching column into a firing line. No indication is given, however, of how close the riflemen of a squad are to be to each other. In contrast, the platoon manual states that in an attack, the four squads of a rifle platoon are to be deployed in a line of 100 m (328 ft) (Sokhan, Shtendera, and Ivanenko 1995:373).

The manual also discusses the role of a sniper and his obligations, which includes "confidently remove a target with one shot, always maintain the weapon and optics in top condition, and use the terrain as a mask" (Sokhan, Shtendera, and Ivanenko 1995:326). In addition, the sniper is to use tracer rounds to provide information about targets such as tanks, emplacements, and artillery that cannot be taken out by rifle fire.

Many UPA documents do not survive, but four available individual inventory documents shed light on UPA combat operations. These documents take the form of a spreadsheet and follow a similar pattern. The first complete document dates to April 2, 1947 (Electronic Archives of the National Liberation Movement [EANLM], Fond 9, Tom 28 2015). The document lists the 42 soldiers of the 3rd Platoon as well as each soldier's nom de guerre, weapon, inventory number, the number of rounds carried, the quantity of grenades (usually one or zero) carried, and the number of rounds fired in battle. This document concludes with a short list of eight soldiers who also carried a pistol.

The first soldier in each squad carried a PPSh-41 Soviet submachine gun, the second soldier was armed with the Degtyaryov (DP) 27 light machine gun, and the remaining members of the squads most often carried rifles or carbines designated as either "Mosin" or "Mauser." No information in this or the following documents discusses what type of "Mosin" or "Mauser" was being used. The listed weapon inventory numbers are intriguing and may refer to the weapons' serial numbers. Given the difficulties in tracking weapon serial numbers from the firearms used by regular, noninsurgent, armies, it is unlikely that this line of inquiry will provide additional information. This document also notes that one soldier, Khmara, was armed with a "sniper."

The second individual inventory document is undated, but one scholar links it to a rifle platoon of the "Udarnyk 4" company and believes that it dates to 1946 (Shchehliuk 2014:146). This document is almost identical to the first, as it also lists each soldier's nom de guerre, weapon, inventory number, number of rounds carried, quantity of grenades (usually one or zero), and number of rounds fired in battle. The weapons listed in this document mirror those noted in the previous inventory, but one soldier carried a "Russian self-loading" weapon. Interestingly, the second document does not include a list of soldiers equipped with pistols, which suggests either that no one carried one or, more likely, that this information was unrecorded. The two additional inventories make no mention of pistols being carried by the UPA soldiers.

The third inventory is incomplete and begins with soldier number 32 (EANLM Fond 9, Tom 28, 4.4.5.2., UPA West 2015). This document is also dated to April 4, 1947. This document includes a partial list of the riflemen of the third squad, all seven of whom were armed with either a Mosin or a Mauser. In place of a light machine gun, one of the members of the fourth squad is noted as being armed with a "Self-loading Mauser."

The last inventory document is also a partial listing (EANLM, Fond 9, Tom 28, 4.4.5.2., UPA West, 2015). This document appears to be the second sheet of the full inventory and begins by listing soldiers number 7 to 10 of what is most likely the 2nd Platoon. It then describes the 10 members of the third squad and concludes with a few additional members of the platoon, including its commander and a medic. Although the document is undated, it notes that four members of the third squad recently died in battle: two on March 4, 1947, another on March 6, and a fourth on March 14. The soldiers in this platoon appear to have followed the same organizational structure as already noted. One soldier in the third squad was equipped with a German assault weapon (Sturmgewehr), and another carried a German self-loading weapon.

The standardized nature of the four documents suggests that they followed an established report format to which platoon leaders were expected to conform. Though not classified as such, these appear to be after-action reports intended for either the company commander or the company logistics officer. According to these four documents, squad organization in the UPA was standardized when possible as well. Although the specifics of each of the inventories showed some variation, each rifle squad had the same general makeup: a squad leader armed with a submachine gun; at times a second or even third man with a submachine gun; one, or in two instances, two light machine gunners who carried a DP; and five to seven riflemen armed with either a Mauser or Mosin rifle. Soldiers armed with self-loading weapons provided additional firepower. The single reference to a soldier armed with a sniper rifle supports the UPA training manual that included a section on snipers.

As the UPA soldiers relied on enemy supplies for armaments, ammunition was a critical nonrenewable resource. With the advance of the Red Army in 1944, new supplies of ammunition for the German and Hungarian weapons with which the insurgents were armed were no longer readily available, and the soldiers had to rely on previously accumulated stores of ammunition until new weapons could be captured. In addition, among the UPA leadership it was commonly believed that a major reason for failure to establish an independent Ukrainian state after World War I was the lack of military supplies and ammunition. Stories were commonly told of Ukrainian soldiers revisiting old trenches in search of small arms ammunition during the 1919 Chortkiv Offensive. Hoping to avoid repeating the mistakes of the previous war, the UPA was keenly aware that all new supplies of weapons and ammunition had to be captured in battle. Therefore, the UPA leadership stressed the need to conserve armaments (Myroslav Symchych, personal communication 2011).

Each of the four inventories shows the number of rounds that each soldier carried and the number expended in a recent engagement. According to the intact April 2, 1947, document, most riflemen carried between 70 and 100 rounds, and a few carried as few as 45. Interestingly, most riflemen did not fire their weapons, and of the eight who did, between 10 and 15 rounds were fired. Submachine gun fire was also limited, with only two of the seven men expending 30 or 60 rounds. Most of the fire was done by the DP light machine guns, which used between 110 and 125 rounds. The single sniper fired five shots. Similar expenditure of ammunition is noted in the second document: the riflemen fired only a few rounds, and the four light machine guns expended between 50 and 100 rounds. The partial document also notes that the riflemen fired infrequently.

In the fourth inventory, all the soldiers, including the riflemen, are listed as having expended much of their ammunition. The light machine gun expended 200 of the 250 rounds carried, and the German Sturmgewehr used 120 of its 150 rounds. The riflemen of the third squad carried between 46 and 90 rounds of ammunition, but by the end of the action they were down to 20 to 30 rounds apiece. Given the high casualty rate suffered by the squad in March 1947, it may not be surprising that they had expended such a high amount of ammunition.

We do not have such after-action reports for the squads of the Birch Company, but the photograph of the entire company provides information as to its strength and possible organization (V'iatrovych et al. 2011). Taken in fall 1944, the photograph shows 111 people standing and six men sitting. Four of the six seated men are armed with submachine guns, and one is armed with a pistol. Note that the photograph also shows a woman in a traditional Ukrainian peasant outfit standing in the first row holding a large bag (Figure 10.5). According to informants, she was a medic who was visiting the camp at the time that the photo was taken.

Figure 10.5. Birch Company, 1944. (Courtesy of Jura Datsko)

Figure 10.6. Close-up view of Birch Company weapons. (Courtesy of Jura Datsko)

Soldiers in the front rank are armed with long-barreled weapons. Six of these show a side mounted, curved box magazine and an extended wooden stock, which suggests that they are armed with Solothurn MG30 light machine guns. One rifle is turned to its side and shows an extended integrated box magazine, two bands, a pronounced rear sight, and an extended unprotected barrel and may be a Hungarian M35/G98/40 (Figure 10.6).

A wartime photograph of Myroslav Symchych shows him armed with a PPS-43 Soviet submachine gun (Halushchak 2013). When asked about his weapon, Symchych stated that he had modified his weapon by replacing the "piece of junk" barrel with a new one he had taken from a recently captured Soviet rifle. He had shortened the barrel by cutting it in half (Myroslav Symchych, personal communication 2011).

Interpretation

As is visible in the map of the ambush site, most identified insurgent firing positions are found midslope on the less steep parts of the mountain ridges. These positions are beneficial for the men setting up the ambush as it places them very close to their target yet still provides a protected firing position. The identified machine gun position is farther away from the bridge where the attack was concentrated and is believed to be one of the two machine

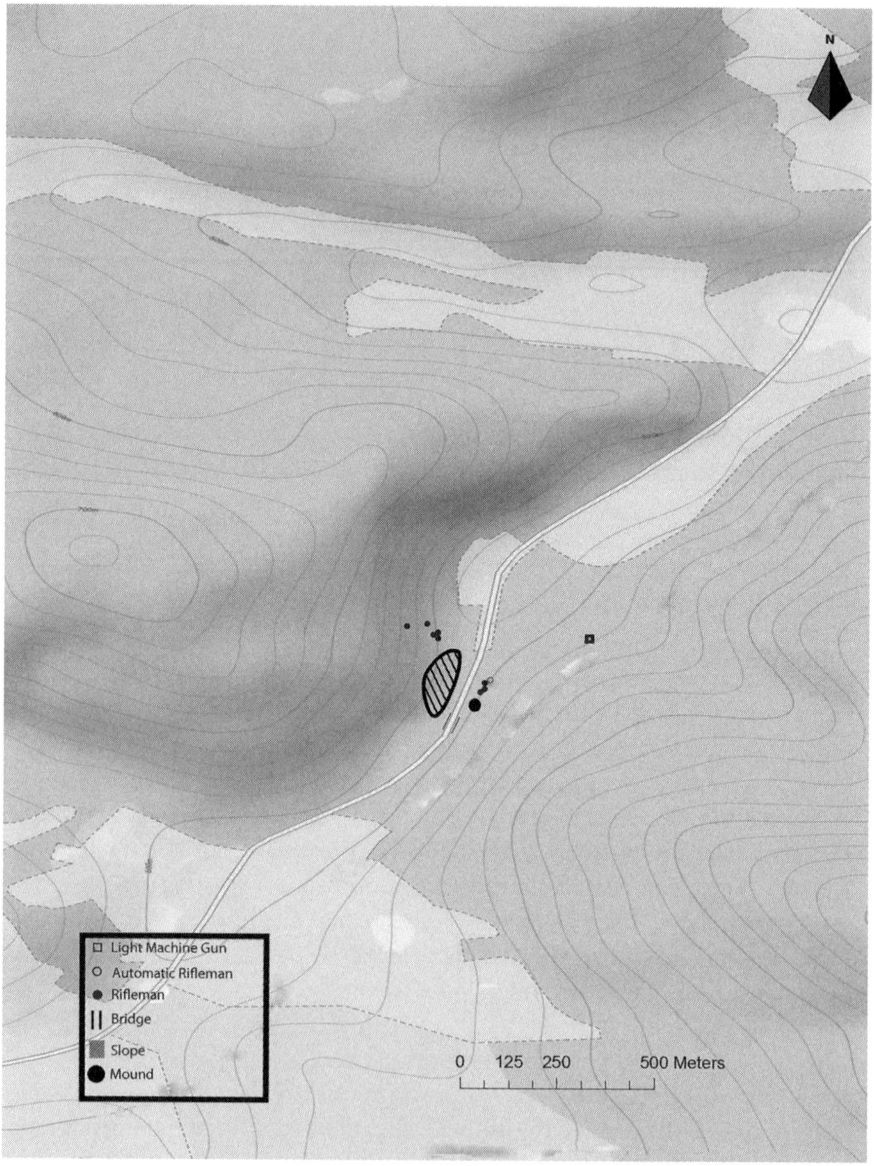

Figure 10.7. Map of ambush site. (Brad Miller, courtesy of Adrian Mandzy)

gun positions set up by Symchych to cover the rear of the enemy column. If so, then the large quantity of expended ammunition illustrates both the effectiveness of this placement and its significance in the engagement with the enemy. The lack of the cartridges that would mark the positions held by individual riflemen near the machine gun suggests that the squad followed the UPA practice of relying on the machine gun for field operations, with the riflemen holding their fire (Figure 10.7).

The firing positions on both the eastern and western slopes also appear to reflect ammunition expenditures and the positions of at least two squads. The UPA soldiers on both slopes were positioned very close to the enemy. In areas where the topography and the geography allowed for artifacts to be preserved, we have evidence that the soldiers in these locations expended a significant amount of ammunition. The best-preserved position, where ammunition clips, the possible remains of a sling, and 39 rounds were recovered, shows the intensity of one rifleman's fire. As the later ammunition reports document, the UPA soldiers carried no more than 100 rounds into battle and often made do with fewer. The cartridges recovered suggest that the position held by this one unknown UPA riflemen was a strong one for firing on the enemy in the valley below.

Other UPA riflemen may have expended less ammunition, or given the size of the Soviet battalion below, fewer cartridges survived to reveal the placement of the troops across the battlefield terrain. The steep nature of the western slope that overlooked the bridge did not facilitate the preservation of any artifacts at that location, and the construction of a memorial on the eastern slope near the bridge destroyed any traces of the position from which Symchych commanded the operation. Undoubtedly, the spring mountain runoff washed away quantities of cartridges, and the construction and rebuilding of the roadway below eradicated traces of the killing zone. Although it is possible that additional fieldwork may uncover more evidence about the ambush site, the terrain, erosion, and construction make it impossible to uncover all the firing positions used by the UPA in their ambush. Despite these limitations, however, our limited multidisciplinary study confirmed the location of the ambush site identified by acting company commander Myroslav Symchych some 70 years after the event. We may never know the name of the UPA soldier from the Birch Company who fired his Mannlicher on the western slope at least 38 times against the enemy below, or that of his brother-in-arms who used his machine gun to cover the tail of the ambush, but this study illustrates what can be learned when applying a multidisciplinary approach to something as simple as a few handfuls of expended Hungarian-manufactured ammunition.

Conclusion

Reflections on the Historical Archaeology of Asymmetric Warfare

Clarence R. Geier

From time immemorial, people made up songs, poems, and stories that were passed from generation to generation perpetuating the myths and legends of great warriors such as Ragnar Lothbrok of Norway, King Arthur of Britain, Boudicca warrior queen of the Iceni, and Achilles of Pythia. Many of these became bigger than life and attained near superhero status, and they were also cultural standard-bearers for the human virtues of courage, honor, dedication, discipline, strength, persistence, and the ability to confront overwhelming odds.

Of the persons and events discussed in this volume, only Frances Marion has attained the level of legend and myth of those mentioned here. Yet groups of individuals joining together under a respected leader to fight perceived injustice, attain political power, change established cultural values, or simply improve their lot in life is common in the human experience. Some of these groups were successful, and some failed. Some are viewed as heroes who shaped history, and others are seen as adversaries or enemies whose self-interest and ideology had to be overcome. All share the circumstance in which a group of lesser power confronted and attempted to overcome a group of greater, usually more established power, hence the idea of asymmetry.

Conflict Archaeology and Historical Archaeology

A defining feature of each of the case studies in this volume is the poor to minimal written record of the events examined. Utilizing the logic of archaeology, we assert that these events leave a material/physical record in the form of artifacts, sets of artifacts, or elements of human construction such as roads, houses, and earthworks. If these items can be recovered, identified, and interpreted with respect to their distribution in space (terrain), an insight into the event that produced them can be obtained.

Just as the historical record is often incomplete, as noted in several of the chapters, the preservation of the archaeological record is often limited and incomplete as well. Much of the material evidence has been lost to natural or human causes in the time since the event under study. These limitations in the historical and archaeological records open the door to the important field of historical archaeology, which combines the research methods and data of history (historiography) and archaeology in the effort to maximize and join the data from each to reveal new facts pertinent to our interpretation of the past.

Given the ongoing role of warfare and the threat of warfare to the present in shaping world politics, a subfield of historical archaeology, variously labeled military, battlefield, or conflict archaeology, has emerged and is frequently a topic addressed in the regular meetings of the Society for Historical Archaeology. Some sessions at these meetings have been turned into successful books (Geier and Potter 2003; Geier and Winter 1995). Further, recent research has contributed to at least two books concentrating on the topics, methods, and theory in the field (Geier et al. 2011; Scott et al. 2007), the development of a regularly offered symposium (Smith 2016), and the establishment of the *Journal of Conflict Archaeology*; graduate programs in the field are also now being offered at prestigious universities. As illustrated in each of the chapters, however, researchers in the field continue to look to the disciplines of history, anthropology, political science, sociology, forensics, and a diversity of natural sciences to evolve a theoretical framework that can be used to better understand the circumstances that mobilize groups of résistance to confront more established and powerful agents of control.

The chapters in this volume illustrate how some of the established techniques of history, such as primary document research (all chapters), genealogical studies (chapter 4), and oral history (chapters 9 and 10) can be used to provide a general to highly specific description of an act of conflict, its participants, location, flow of action over time and space, and outcomes. A variety of archaeological methods are discussed, which, in conjunction with historical accounts, are used to either create a narrative of a military action or verify the stages of its progress. Many of the studies use archaeological testing and excavation (chapters 1, 3, 4, and 9) and strategies of metal detecting as sources of artifact data (chapters 1, 2, 3, 4, 6, 8, 9, and 10). These chapters discussing metal detecting also illustrate the fundamental importance of applying forensic and other more traditional techniques, first to identify event-associated artifacts and their status (fired vs. unfired bullets, etc.) and then to analyze their placement in their natural and human-shaped settings. Analysis of both natural and cultural landscapes believed in existence at the time are considered as they contributed to the event.

KOCOA and Its Uses

Practically all chapters discuss a visual reconnaissance and terrain or landscape analysis method known by its acronym KOCOA, explained in the introduction. As pioneered by the National Park Service American Battlefield Protection Program, the method is a process by which critical terrain features are identified and evaluated for their impact on opposing forces in a battle situation. The park service originally saw this method, or planning tool, as a means to discover and precisely define the boundaries of battlefields lost in history. As such, it was applied at the tactical level. This volume, however, clearly demonstrates its multifunctional utility.

Sivilich (chapter 5) in the discussion of the Second Seminole War (1835–1842) and the McBrides (chapter 1) in their discussion of Native American raids on encroaching European settlement during the French and Indian War, both challenge a "one size fits all" application of the KOCOA approach. When dealing with opposing forces from different cultures, the manner in which they understand, interpret, envision, and utilize terrain in confronting their adversary will differ. Military or war bands of different ethnicities will approach their enemy differently when attacking and will respond differently when under attack. When applying a KOCOA approach in the analysis of such engagements, one needs an accurate knowledge of the ability of each to wage war and of their established traditions of conflict.

The above logic is relatively clear when dealing with military forces of different cultures; however, the same can be true when dealing with asymmetric or irregular conflicts between groups of similar cultural identity and origin as illustrated in Smith's discussion of the partisans led by Francis Marion in South Carolina during the American Revolution (chapter 4). Early in the war these rebel bands, confronted with limited manpower and supply, fought a very different war against the British forces of the king as compared with the later war, when the tide began to turn against the British, and Marion's partisans cooperated with the colonial army.

A significant factor in the analysis of any military action, regardless of whether it includes conflict, is its scope, as measured both in its duration and in the physical dimensions of the landscape it comes to encompass. Here again, the KOCOA approach has much to offer. A student of virtually any battle (Antietam, Gettysburg, Cowpens, etc.) knows that each such action evolves and takes on a life of its own. Over the action's duration, groups of diverse sizes and technical capability (infantry, cavalry, artillery, etc.) hold a position, move, or are moved, as desired or as possible, to attain one or more favorable outcomes. In a nonconflict example, the scope of the partisan community supporting Francis Marion was made up of spatially, socially, functionally, and ecologically diverse elements, which morphed in shape and viability in response to British intrusions and disruptions over time. As illustrated in this volume, the KOCOA approach can be used to elucidate and identify the features that shape the interpretation of typically complex and expansive battlefield landscapes as well as the contextual features of specific, smaller actions or engagements that make up the larger event. Once the core of the battlefield is generally located, KOCOA allows the archaeologist to work outward, to identify troop movements and other aspects of the battlefield, commonly termed *defining features* by the American Battlefield Protection Program (chapters 3, 5, and 6).

Technological Adaptation in Asymmetric Conflict

As many twentieth-century conflicts (such as that of the US military in Vietnam and the Russians in Afghanistan) have demonstrated, superior technology does not automatically translate into military victory on the battlefield. Haecker (chapter 8) illustrates this point by discussing how the perceived advantage of the technologically "superior" Spanish harquebusiers proved erroneous when the single shot fired by that weapon was measured against the number of arrows that could be fired in the time it took to reload. He also discusses how Spanish control of La Gran Chichimeca waned because of increased Apache raids, bringing to question the issue of "superiority" of force. On their turf and on their terms, the Apache warriors were superior to the "technologically advanced" Spanish. As Haecker notes, this circumstance was inherited by the United States when the Apache lands of what would become New Mexico and Arizona were acquired after the Mexican-American War. This became a particularly significant issue during the American Civil War when the Apache leaders Mangas Coloradas and Cochise attempted to take advantage of the diminished military force on the frontier.

Likewise, history provides numerous examples of indigenous tribal peoples adapting more quickly to technological change than do nation-states. From the onset of European efforts to settle and exploit the resources of the New World, examples abound. European market demands in the infancy of the colonial experience had a profound impact on changing and shaping the power structure of Native American tribes in the Northeast, and technology provided the means. Driven by a powerful European market in furs (soft gold), particularly beaver pelts, the Dutch (1610–1614), and subsequently the English, entered into trade relations with the Iroquois League of Five Nations (Seneca, Cayuga, Onondaga, Oneida, Mohawk) for access to this highly profitable commodity. In an effort to stimulate trade, the Dutch provided members of the league with guns, gunpowder, and other European goods to develop their hunting and fur-trapping skills. Instead, access to this new military technology combined with the political advantage provided by their alliance, mobilized the Iroquois League in a way that changed the balance of power between them and their Iroquois and Algonquin neighbors. By the second quarter of the seventeenth century, the confederacy undertook to expand its territory and monopolize the fur trade between the European markets and the tribes of the Great Lakes. Through a series of brutal and bloody conflicts fought across the seventeenth century, the league engaged in actions that established its unprecedented power and also reshaped the Native American tribal landscape in the Northeast. For a time the league was one of the most powerful political forces in North America, having virtual control of the Ohio Valley west to the Mississippi and into the Lower Peninsula of Michigan (Hunt 1960; Jennings 1984; Kehoe 2006: 227–244; Trigger 1976). The combined occurrence of a working Native American confederation, an alliance with Dutch and then English market interests, and the effective incorporation of a new "super weapon" (guns) changed what had been a relatively equal and balanced power between enemies.

There was a major impact on the indigenous tribes. In response to European influence, the Iroquois gradually changed their reasons for fighting. Whereas war was initially fought for prestige, revenge, or acquisition of goods or captives, their motive shifted to one of economic advantage and control over the bountiful beaver hunting grounds of the Great Lakes and Ohio Valley (Hall 2017).

Asymmetric Warfare and Primitive Warfare

The reliance on stealth, surprise, and ambush as principal tactics of asymmetric war used by the Iroquois is significant in that these are the primary tactics of the less powerful parties of asymmetric conflict (Malone 1991). In this book, this is illustrated in the behavior of Ukrainian insurgents (chapter 10), the Apache (chapter 8), the Shawnee and other tribes that mobilized to attack encroaching European settlements in the French and Indian War (chapter 1), the raids of Francis Marion's men against the British (chapter 4), and the Seminole war bands that fought against the US military (chapter 5). As Smith points out in the introduction, these tactics are the prime tactics of primitive warfare providing the aggressor group the greatest advantage of success with the least risk.

Options for Insurgents in Asymmetric Warfare

Smith (chapter 4) stresses the importance of a politically aligned and supportive community to the security and viability of an insurgency. Such is not always the case, however. Another option for the security of an insurgency is aggressive terrorism to subdue all resistance. The use of such terror tactics to create a relatively safe working environment for a revolutionary group can be seen in the recent actions of the Islamic State of Iraq (ISIS, ISIL) and Al Qaeda, who use mass murder, imprisonment, rape, and the threat of them to terrorize and neutralize resistance in their areas of control (Fleischman 2015). Similarly, in the United States and elsewhere, the international criminal gang MS-13 establishes its power and control in a community through a notorious use of violence and a moral code that predominantly consists of merciless revenge and cruel retribution (Adams and Pizarro 2009; Logan 2009; United States Government 2017).

Many groups, however, when confronted by resistance from stronger groups in their home societies, chose to leave their homelands and search for opportunity in far-off lands. This response is sometimes matched by an effort of the larger community to either drive them out or otherwise eradicate them. This fact contributed to European settlement of the Americas by groups such as the Mennonites, Huguenots or French Protestants, and Puritans (Pennsylvania), who sought freedom from religious persecution in their homelands. Another example of this is the little-known Mormon War of 1838 (Missouri Mormon War), the subsequent events of Mormon persecution, and ultimately the relocation of the faithful to Utah (Greene 1839; Hafen and Hafen 1958; Nauvoo Tourism Office 2017; Poll and Hansen 1961; Rawson and Lyman 2014).

Asymmetric Warfare and Vigilantism

Drexler's discussion of the unrest on the Confederate home front in southwest Arkansas (chapter 7) draws attention to nonconventional warfare and insurgencies that emerged in many of the marginal areas of the United States and the Confederacy during the Civil War. Scott (chapter 6) discusses issues with the Jayhawkers, guerrilla fighters (some say "bandits") associated with Kansas Free-Staters who often clashed with proslavery groups from Missouri. Drexler introduced the idea of lay-out gangs, a different sort of resistance generated by groups motivated to resist the policies of the Confederate government. These groups could be made up of men with pro-Union beliefs who sought to disrupt local Confederate actions. Others were draft dodgers—anti-Confederate because of policies such as impressment and conscription—or deserters who took advantage of the chaos of the time to resist Confederate authorities. Some of the groups lived at the margin of society, becoming bandits who harassed local citizens and military units for their own personal gain (Williams 2010). Indeed, asymmetric warfare provides fertile ground for the growth of group-on-group violence under the cover of state-level warfare. Thus, where conventional warfare occurs, groups variously called lay-out gangs, vigilantes, banditti, and bushwhackers inevitably emerge.

Bushwhacking, a form of guerrilla warfare involving nonuniformed, irregular military forces, was active on both sides of the American Civil War. Some bushwhacking groups even operated with the support and approval of established governments (see Drexler, chapter 7). They usually emerged in large areas of contested land where the lack of available government resources prevented effective control and where there were sharp divisions in community allegiance. Some bushwhackers conducted well-organized raids against the military, but the attacks of others involved ambushes on individuals or families in rural areas and often amounted to fighting between neighbors to settle personal accounts (Kozikowski 2017; O'Bryan 2017).

While certain of the "bushwhackers" such as Bloody Bill Anderson and William Quantrill became infamous for their cruelty, insurgent groups of "partisan rangers" in Virginia's Shenandoah Valley played a key role in mitigating the influence and control of the Union forces in the valley and east of the Blue Ridge Mountains. McNeill's Rangers, an independent Confederate military force (Company E, 18th Virginia Cavalry) commissioned under the Partisan Ranger Act of 1862, was one of two such groups allowed to remain active after the repeal of the act in 1864 (Sutherland 2015). John Hanson "Hanse" McNeill's successful and frequent raids on the B&O Railroad in West Virginia are believed to have resulted in the diversion of significant numbers of Union troops from other service in the eastern theater, to protect the critical rail line (Delauter 1986; Rada 2013).

And then there was the "grey ghost," John Singleton Mosby and the 43rd Battalion, Virginia Cavalry, known as Mosby's Rangers or Raiders. Mosby's Rangers was the second of the partisan ranger groups to be authorized by the Confederate government to function in Virginia after 1864. Known for their strategic strikes on Union targets and their ability to blend into the local citizenry to escape Union pursuers, Mosby's Rangers had an area of operations that encompassed northern Virginia between the Shenandoah Valley and Alexandria, in what came to be known as "Mosby's Confederacy." In the history of the Shenandoah Valley, Mosby is best known for his contest with Major General Philip Sheridan, playing havoc with Sheridan's lines of supply during the 1864 Valley Campaigns (Evans and Moyer 1991; Mosby 2006; Ramage 2009).

As noted, the Civil War provided a cover, or legitimacy, and opportunity for acts of revenge against neighbors. Smith's chapter on Marion's partisan community (chapter 4) and its conflict with nearby Loyalist communities demonstrates how these feuds go deep into history. Indeed, these feuds have their roots in the Old World (Fischer 1989; McWhiney and Jamieson 1984:171–191). McBride (chapter 9) brings them closer to the present in one of the legendary post–Civil War events in the East: the Hatfield and McCoy feud (1863–1891) in the area of the Tug Fork of the Big Sandy River on the border of West Virginia and Kentucky. McBride demonstrates the connection between Civil War–period western bushwhackers (Drexler, chapter 7) and the eastern partisan, guerilla raids of the Civil War to later Appalachian feuding. While the feud was not at a level that would shape North America's historical trajectory, America's fascination with the story has made it the subject of stories, movies, and its own three-part miniseries produced by the History Channel in 2012. The story is significant because it addresses the human reality in many parts of the United States following the Civil War, where, even though the actual war was over, the animosity between families in areas of divergent allegiance often continued. The McBride chapter also brings out the point that asymmetric conflicts do not have to occur at a state or pancultural level to have an impact on human lives, nor do they have to involve trained and organized military units.

Mandzy (chapter 10) draws attention to the plight of certain ethnic and nationalistic groups in Europe and elsewhere before, during, and following World War II. While the United States was not active in supporting the Ukrainian movement, the role of Uncle Sam and other Allies in supporting insurgent actions during the war was pivotal to success in many areas: the Ramon Magsaysay guerillas in the Philippines, the Norwegian resistance, Wingate's Chindits in Burma, La Résistance in France, and the anti-Soviet Forest Brothers of Latvia, Lithuania, and Estonia. From the other side, anti-American insurgent groups have become a source of threat on the world scene: the Viet Cong in Vietnam, the Taliban in Afghanistan, and the Hezbollah in Lebanon. Twenty-first century insurgents, including the most dangerous, have no direct state ties but instead have strong religious affiliation, as seen with Al Qaeda and ISIS.

Concluding Thoughts

The preceding chapters seek to contribute to a developing working theory and historical-archaeological methodology that can be applied to document and interpret the conflicts that make up the history of many of these movements and that are often absent from traditional historical accounts. Although most of the chapters address what may be seen as relatively small events, what is significant is that when combined with other such events over time and space, they can result in great events. The rebels of the South Carolina swamps joined in a struggle that became the United States of America, and the persecuted members of the Ukrainian resistance persisted and ultimately attained an independent Ukraine with the dissolution of the Soviet Union in 1991.

References Cited

Adams, Christopher D.

2012 New Perspectives on the Apache Frontier in Nuevo Mexico: 16th, 17th and 18th Century Apache Sites along the Seco Creek, Black Range Mountains, New Mexico. Poster presented at the Pecos Conference, Pecos National Park, New Mexico.

Adams, Christopher D., Diane E. White, and David M. Johnson

2000 *Last Chance Canyon 1869 Apache/Cavalry Battle Site*. Lincoln National Forest, Alamogordo, New Mexico.

Adams, Jennifer J., and Jesenia M. Pizarro

2009 MS-13: A Gang Profile. *Journal of Gang Research* 16(4):1–14.

Alther, Lisa

2012 *Blood Feud: The Hatfields and the McCoys: The Epic Story of Murder and Vengeance*. Lyons Press, Guilford, Connecticut.

Arbuckle, Matthew

1778 Letter to Gen. Edward Hand, June 2. Draper Manuscripts 18J76. State Historical Society of Wisconsin, Madison. (Reprinted in Kellogg 1916: 64–65.)

Arensberg, Conrad M., and Solon T. Kimball

1965 *Culture and Community*. Harcourt, Brace and World, New York.

Army and Navy Chronicle [Washington, DC]

1836 Report on the Peninsula War. 19 May. 3:318.

Asprey, Robert B.

1975 *War in the Shadows: The Guerrilla in History*. 2 vols. Doubleday, Garden City, New York.

Babits, Lawrence E.

1998 *A Devil of a Whipping: The Battle of Cowpens*. University of North Carolina Press, Chapel Hill.

2004 Asymmetric Warfare during the American Revolution's Southern Campaign. Paper presented at the Annual Meeting of the Society for Military History, Charleston, South Carolina.

2010 Patterning in Earthen Fortifications. In *Historical Archaeology of Mili-*

tary Sites, edited by Clarence E. Geier, Lawrence E. Babits, Douglas D. Scott, and David G. Orr, pp. 113–121. Texas A&M University Press, College Station.

2014 Appendix: METT-T, KOCOA, and the Principles of War: A Template Guiding a Better Understanding of Battlefield Behavior and Detritus. In *From These Honored Dead: Historical Archaeology of the American Civil War*, edited by Clarence R. Geier, Douglas D. Scott, and Lawrence E. Babits, pp. 263–270. University Press of Florida, Gainesville.

Bailey, Anne J., and Daniel E. Sutherland (editors)

2000 *Civil War Arkansas: Beyond Battles and Leaders*. University of Arkansas Press, Fayetteville.

Bailey, DeWitt

2002 *British Military Flintlock Rifles, 1740–1840*. Andrew Mowbray, Lincoln, Rhode Island.

Ball, Eve

1970 *In the Days of Victorio: Recollections of a Warm Springs Apache*. University of Arizona Press, Tucson.

Barnes, Frank C.

2000 *Cartridges of the World: A Complete and Illustrated Reference for over 1500 of the World's Most Important Sporting Cartridges*. Edited by M. L. McPherson. 9th ed. Krause, Iola, Wisconsin.

Barnes, Kenneth C.

2001 The Williams Clan's Civil War: How an Arkansas Farm Family Became a Guerilla Band. In *Enemies of the Country: New Perspectives on Unionism in the Civil War South*, edited by John C. Inscoe and Robert C. Kenzer, pp. 188–207. University of Georgia Press, Athens.

Barnett, Roger W.

2003 *Asymmetrical Warfare: Today's Challenge to U.S. Military Power*. Brassey's, Washington, DC.

Bass, Robert D.

1959 *Swamp Fox: The Life and Campaigns of General Francis Marion*. Alvin Redman, London.

Basso, Keith (editor)

1971 *Western Apache Raiding and Warfare from the Notes of Grenville Goodwin*. University of Arizona Press, Tucson.

Baum, Dale

2008 Slaves Taken to Texas for Safekeeping during the Civil War. In *The Fate of Texas: The Civil War and the Lone Star State*, edited by Charles D. Grear, pp. 83–104. University of Arkansas Press, Fayetteville.

Beers, F. W.

1873 *Atlas of Douglas County, Kansas*. F. W. Beers, New York.

Betros, Lance

2004 *West Point: Two Centuries and Beyond*. Texas A&M University Press, College Station.

Bickley, Ancella R.

1993 Searching for the Historical Dick Pointer. *Journal of the Greenbrier Historical Society* 6(2):16–24.

Billings, Dwight B., and Kathleen M. Blee

2000 *The Road to Poverty: The Making of Wealth and Hardship in Appalachia.* Cambridge University Press, New York.

Birdwell, Arthur

1999 John Brown's First Battle: A Brief History of Sites in the Black Jack Area in Douglas County, complied by Dorthea I. Jackson. Manuscript on file, Baldwin City Public Library, Baldwin City, Kansas.

Birtle, Andrew J.

2003 The Origins of the Legion of the United States. *Journal of Military History* 67(4):1249–1262.

Blee, Kathleen M., and Dwight B. Billings

1996 Violence and Local State Formation: A Longitudinal Case Study of Appalachian Feuding. *Law and Society Review* 30(4):671–706.

Bletzer, Michael P.

2016 Down the Mountain, Up the Mountain: Peace and War on the Chichimeca Frontiers, 1590–1800. *Papers of the Archaeological Society of New Mexico* 42:11–22.

Blight, David W.

2001 *Race and Reunion: The Civil War in American Memory.* Belknap Press, Cambridge, Massachusetts.

Board of Visitors

1821 Annual Report of the Board of Visitors to the United States Military Academy Made to Congress and the Secretary of War for the Year 1821. West Point, New York.

Boatner, Mark M.

1966 *Encyclopedia of the American Revolution.* David McKay, New York.

Bol'novs'kyj, Andrij

2003 *Ukrains'ki vijs'kovi formuvannia v zbrojnykh sylakh Nimechchyny (1939–1945)* [Ukrainian military formations in the German Armed Forces (1939–1945)]. Ivan Franko L'viv. National University Press, L'viv, Ukraine.

Bondi, August

1904 With John Brown in Kansas. *Transactions of the Kansas State Historical Society 1903–1904* 8:281–289.

Bonfig, Audrey E.

2016 The Hatfields and McCoys: America's Feud, as Portrayed in the New York, West Virginia, and Kentucky Press, 1888–1890. Master's thesis, School of Journalism, Ohio University, Athens.

Borovych, Ia. V.

1986 Ukraine and Poland. Reprinted in *Political Thought of the Ukrainian Underground, 1943–1951*, edited by Peter J. Potichnyj and Yevhen Shtendera, pp. 57–68. Canadian Institute of Ukrainian Studies Press, Toronto. Originally published 1943, *Ideia i Chyn* 2(4).

Bradshaw, John

1833–1835 Revolutionary War Pension Application S6738. National Archives and Records Administration, Washington, DC.

Brumwell, Stephen

2002 *Redcoats: The British Soldier and War in the Americas, 1755–1763*. Cambridge University Press, Cambridge.

Buchanan, John

1997 *The Road to Guilford Courthouse: The American Revolution in the Carolinas*. John Wiley and Sons, New York.

Buffaloe, David L.

2006 *Defining Asymmetric Warfare*. Land Warfare Papers No. 58. Institute of Land Warfare, Association of the United States Army, Arlington, Virginia.

Burgoyne, Bruce E. (translator)

1987 Journal Kept by the Distinguished Hessian Field Jäger Corps during the Campaign of the Royal Army of Great Britain in North America. *Journal of the Johannes Schwalm Historical Association* 3(3):45–62.

1995 The Hesse-Cassel Field Jäger Corps—"Das Fieldjäger Corps von Hessen-Kasse im Amerikanschen Unabhaegigkeitskrieg, 1776–1783." *Journal of the Johannes Schwalm Historical Association* 5(3):8–10.

1997 *Diaries of Two Anspach Jägers: Lieutenant Heinrich Carl Phillip von Feilitzsch and Lieutenant Christian Friedrich Bartholomai*. Heritage Books, Bowie, Maryland.

Butler, David Stewart

2001 *An Archaeological Model of Seminole Combat Behavior*. Master's thesis, Department of Applied Anthropology, University of South Florida, Tampa.

Butler, Lindley

1994 David Fanning's Militia: A Roving Partisan Community. In *Loyalist and Community in North America*, edited by Robert M. Calhoon, Timothy M. Barnes, and George A. Rawlyk, pp. 147–157. Greenwood Press, Westport, Connecticut.

Butler, Scott

2011 *Metal Detector Survey and Battlefield Delineation of the Buford's Massacre (Waxhaws) Revolutionary War Battlefield, SC Route 9 and Route 522, Lancaster County, South Carolina*. Brockington and Associates. Report prepared for the South Carolina Department of Transportation, Columbia.

Calloway, Colin G.

1997 *New Worlds for All: Indians, Europeans, and the Remaking of Early America*. Johns Hopkins University Press, Baltimore, Maryland.

2007 *The Shawnee and the War for America*. Penguin Group, New York.

Cartmill, Thomas

ca. 1840 Interviews by the Rev. Dabney Shane. Draper Manuscripts 12CC61. State Historical Society of Wisconsin, Madison.

Casdorph, Paul D
1996 *Prince John Magruder: His Life and Campaigns*. John Wiley and Sons, New York.

Catts, Wade P., Juliette Gerhardt, Elisabeth LaVigne, Peter Leach, and Timothy Mancl
2014 *"An Opportunity of Annoying Them Greatly": Archaeological Investigation of a Portion of the Cooch's Bridge Battlefield*. John Milner Associates. Report prepared for Delaware Division of Historical and Cultural Affairs, Dover.

Catts, Wade P., Timophy Mancl, and Elizabeth Lavigne
2012 *Report of a Phase I Archaeological Survey of the Proposed Retention Basin, Sunset Lake, New Castle County, Delaware*. John Milner Associates. Report prepared for the New Castle Conservation District, Newark, Delaware.

Catts, Wade P., Robert A. Selig, Elisabeth LaVigne, Kevin Bradley, Kathryn Wood, and David G. Orr
2017 *"It is Painful for me to Lose so Many Good People": Report of an Archeological Survey at Red Bank Battlefield Park (Fort Mercer), National Park, Gloucester County, New Jersey*. Commonwealth Heritage Group. Report prepared for the Gloucester County Department of Parks and Recreation, Clarksboro, New Jersey.

Chacon, Richard J., and Michael C. Scoggins
2014 *The Great Awakening and Southern Backcountry Revolutionaries*. Springer Press, New York.

Chang, K. C.
1968 Toward a Science of Prehistoric Society. In *Settlement Archaeology*, edited by K. C. Chang, pp. 1–9. National Press Books, Pal Alto, California.

Coe, Charles H.
2016 [1898] *Red Patriots: The Story of the Seminoles*. Editor, Cincinnati, Ohio. 2016 facsimile ed. Applewood Books, Carlisle, Massachusetts.

Cohen, M. M.
2016 [1836] *Notices of Florida and the Campaigns*. Burgess and Honour, New York. 2016 facsimile ed. Applewood Books, Carlisle, Massachusetts.

Collett, Chris
2007 *Counteracting Irregular Activity in Civil War Arkansas—A Case Study*. United States Army War College, Carlisle Barracks, Pennsylvania.

Collins, James Potter
1859 *Autobiography of a Revolutionary Soldier*. Edited by John M. Robert. Feliciana Democrat, Clinton, Louisiana.

Connelly, Thomas
1964 Indian Warfare on the Tennessee Frontier, 1776–1794: Strategy and Tactics. *East Tennessee Historical Society* 36:3–22.

Connor, Melissa, and Douglas D. Scott
1998 Metal Detector Use in Archaeology: An Introduction. *Historical Archaeology* 32(4):73–82.

Cooch, E. W.
1940 *The Battle of Cooch's Bridge*. Privately published, Cooch's Bridge, Delaware.
Craig, Alan K., and Christopher Peebles
1974 Ethnoecologic Change among the Seminoles, 1740–1840. *Geoscience and Man* 5:83–96.
Craig, John
1854 The War in York and Chester. *Chester Standard*. 16 March. Chester, South Carolina.
Crawford, B. Scott
2008 A Frontier of Fear: Terrorism and Social Tension along Virginia's Western Waters, 1742–1775. *West Virginia History*, New Series 2(2):1–29.
Crawford, T. C.
1889 *An American Vendetta: A Story of Barbarism in the United States*. Belford, Clarke, New York.
Cullum, George W.
1868 *Biographical Register of the U.S. Military Academy, at West Point, N.Y., from Its Establishment March 16, 1802 to the Army Re-Organization of 1866–1867*. D. Van Nostrand, New York.
Cummings, Andrew
1832 Pension Application W911. Record Group M804, National Archives and Records Administration, Washington, DC.
Dansey, William
1777 Papers. Delaware Historical Society, Wilmington, Delaware.
Davie, William Richardson
1976 *The Revolutionary War Sketches of William R. Davie*. Edited by Blackwell P. Robinson. North Carolina Department of Cultural Resources, Division of Archives and History, Raleigh.
Dayton, Ruth
1942 *Greenbrier Pioneers and Their Homes*. West Virginia Publishing, Charleston.
Dean, Nadia
2012 *A Demand of Blood: The Cherokee War of 1776*. Valley River Press, Cherokee, North Carolina.
Dederer, John
1983 *Making Bricks without Straw: Nathanael Greene's Southern Campaigns and Mao Tse Tung's Mobile War*. Sunflower University Press, Manhattan, Kansas.
Delauter, Roger U., Jr.
1986 *McNeill's Rangers*. 2nd ed. Virginia Regimental Histories Series. H. E. Howard, Lynchburg, Virginia.
Dennis, Jeff W.
2017 *Patriots and Indians: Shaping Identity in Eighteenth-Century South Carolina*. University of South Carolina Press, Columbia.
Department of the Army
2008 *Field Manual 3-0: Operations*. Department of the Army, Washington, DC.

Díaz del Castillo, Bernal
1956 [1632] *Discovery and Conquest of Mexico*. Edited by Genaro Garcia. Farrar, Straus and Cudahy, New York.

Division of Historical Resources
2015 *Florida Seminole Wars Heritage Trail Guide*. Florida Department of State, Tallahassee. http://dos.myflorida.com/media/695430/seminole_war_heritage_trail.pdf, accessed August 8, 2018.

Doddridge, Joseph
1824 *Notes on the Settlement and Indian Wars*. Privately published, Wellsburg, Virginia.

Dolphin, Richard
1977 *The International Book of Beer Can Collecting*. Castle Books, Secaucus, New Jersey.

Dotson, Tom E.
2014 *The Hatfield and McCoy Feud after Kevin Costner: Rescuing History*. CreateSpace, North Charleston, South Carolina.
2015a *Crazy Jim Vance and the Men Who Murdered Him*. CreateSpace, North Charleston, South Carolina.
2015b *The Missing McCoys*. CreateSpace, North Charleston, South Carolina.

Dougan, Michael B.
1976 *Confederate Arkansas: The People and Policies of a Frontier State in Wartime*. University of Alabama Press, Tuscaloosa.

Downes, Randolph C.
1968 *Council Fires on the Ohio*. University of Pittsburg Press, Pittsburgh, Pennsylvania.

Drexler, Carl G.
2013 Dooley's Ferry: The Archaeology of a Civilian Community in Wartime. PhD dissertation, Department of Anthropology, College of William & Mary, Williamsburg, Virginia.

Dye, David
2006 The Transformation of Mississippian Warfare: Four Case Studies from the Mid-South. In *The Archaeology of Warfare: Prehistories of Raiding and Conquest*, edited by Elizabeth N. Arkush and Mark W. Allen, pp. 101–147. University Press of Florida, Gainesville.

Dyer, Frederick H.
1908 *A Compendium of the War of the Rebellion*. Dyer, Des Moines, Iowa.

Edgar, Walter B.
2001 *Partisans and Redcoats: The Southern Conflict That Turned the Tide of the American Revolution*. William Morrow, New York.

Eid, Leroy V.
1988a "A Kind of Running Fight": Indian Battlefield Tactics in the Late Eighteenth Century. *Western Pennsylvania Historical Magazine* 71(2):147–171.
1988b Their Rules of War: James Smith's Summary of Indian Woodland War. In *Selected Papers from the 1985 and 1986 George Rogers Clark Trans-*

Appalachian Frontier History Conferences, edited by Robert J. Holden, pp. 63–80. Vincennes University, Vincennes, Indiana.

Espenshade, Christopher T., Robert L. Jolley, and James B. Legg
2002 The Value and Treatment of Civil War Military Sites. *North American Archaeologist* 23(1):39–67.

Evans, Thomas J., and James M. Moyer
1991 *Mosby's Confederacy: A Guide to the Roads and Sites of Colonel John Singleton Mosby*. White Mane, Shippensburg, Pennsylvania.

Ewald, Johann
1979 *Diary of the American War: A Hessian Journal, Captain Johann Ewald, Field Jäger Corps*. Edited and translated by Joseph P. Tustin. Yale University Press, New Haven, Connecticut.

Fairbanks, Charles H.
1974 *Florida Indians: III. Ethnohistorical Report on the Florida Indians*. Garland, New York.

Ferguson, Clyde R.
1979 Functions of the Partisan-Militia in the South during the American Revolution: An Interpretation. In *The Revolutionary War in the South: Power, Conflict, and Leadership*, edited by W. Robert Higgins, pp. 239–258. Duke University Press, Durham, North Carolina.

Ferguson, R. Brian, and Neil L. Whitehead
1992 The Violent Edge of Empire. In *War in the Tribal Zone*, edited by R. Brian Ferguson and Neil L. Whitehead, pp. 1–30. School of American Research Press, Santa Fe, New Mexico.

Ferling, John E.
1980 *A Wilderness of Miseries: War and Warriors in Early America*. Greenwood Press, Westport, Connecticut.

Fischer, David Hackett
1989 *Albion's Seed: Four British Folkways in America*. Oxford University Press, New York.
1994 *Paul Revere's Ride*. Oxford University Press, New York.

Fitz-Simons, Daniel E.
1995 Francis Marion the "Swamp Fox": An Anatomy of a Low-Intensity Conflict. *Small Wars and Insurgencies* 6(1):1–16.

Fleischman, Julian
2015 Fear Tactics Have Proven Successful for ISIS. Center for Security Policy, June 26, 2015. www.centerforsecuritypolicy.org/2015/06/26/fear-tactics-have-proven-successful-for-isis/, accessed June 15, 2017.

Foard, Glen, and Richard Morris
2012 *The Archaeology of English Battlefields: Conflicts in the Pre-Industrial Landscape*. CBA Research Report No. 168. Council for British Archaeology, St Mary's House, York, United Kingdom.

Fountain, Albert J.
1962 The Battle of Apache Pass. In *Arizona Cavalcade: The Turbulent Times*, edited by Joseph Miller, pp. 31–35, Hastings House, New York.

Gardner, Mark L.
2000 *Wagons for the Santa Fe Trade: Wheeled Vehicles and Their Makers, 1822–1880*. University of New Mexico Press, Albuquerque.

Gat, Azar
1999 The Pattern of Fighting in Simple, Small-Scale, Pre-State Societies. *Journal of Anthropological Research* 55(4):563–583.
2000a The Human Motivational Complex: Evolutionary Theory and the Causes of Hunter Gatherer Fighting. Part I. Primary Somatic and Reproductive Causes. *Anthropological Quarterly* 73(1):20–34.
2000b Part II: Proximate, Subordinate, and Derivative Causes. *Anthropological Quarterly* 73(2):78–88.

Geier, Clarence R., and Stephen Potter
2003 *Archaeological Perspectives on the American Civil War*. University Press of Florida, Gainesville.

Geier, Clarence R., Douglas D. Scott, and Lawrence E. Babits
2011 *From These Honored Dead: Historical Archaeology of the American Civil War*. University Press of Florida, Gainesville.

Geier, Clarence R., and Susan Winter
1995 *Look to the Earth: Historical Archaeology of the American Civil War*. University of Tennessee Press, Knoxville.

Gillilan, James
1833–1835 Revolutionary War Pension Application R4029. National Archives and Records Administration, Washington, DC.

Gluckman, Arcadi
1965 *Identifying Old U.S. Muskets, Rifles, and Carbines*. Bonanza Books, New York.

Graham, David
1899 *History of the Graham Family*. Privately published, Clayton, West Virginia.
1905 Indian Raid on the Home of Col. James Graham, Greenbrier River, 1777. *West Virginia Historical Magazine Quarterly* 5(1):13–17.

Gray, Robert
1780 Letter to Lord Charles Cornwallis, September 30. Cornwallis Papers, PRO 30/11/64:130–131. National Archives, London.
1910 Colonel Robert Gray's Observations on the War in Carolina. *South Carolina Historical and Genealogical Magazine* 11(3):139–159.

Greene, John P.
1839 *Facts Relative to the Expulsion of the Mormons or Latter-Day Saints, from the State of Missouri, under the "Exterminating Order."* R. P. Brooks, Cincinnati, Ohio.

Greene, Nathanael
1980 [1777] General Greene's Orders, Camp at White Clay Creek, August 28, 1777. In *The Papers of General Nathanael Greene: II. January 1777–October 1778*, edited by Richard K. Showman, Robert E. McCarthy, and Margaret Cobb, p. 147. University of North Carolina Press, Chapel Hill.

1991 [1780] Letter to Francis Marion, December 4, 1780. In *The Papers of General Nathanael Greene: VI. 1 June 1780–25 December 1780*, edited by Richard K. Showman, Dennis M. Conrad, Roger N. Parks, and Elizabeth C. Stevens, pp. 519–520. University of North Carolina Press, Chapel Hill.

Gridley, Karl L.

2001 John Brown and Lawrence, Kansas Territory, 1855–1859. In *Embattled Lawrence: Conflict and Community*, edited by Dennis Domer and Barbara Watkins, pp. 23–39. University of Kansas Continuing Education, Lawrence.

Griffith, G. W. E.

1925 The Battle of Black Jack. *Collections of the Kansas State Historical Society* 16:524–528.

Griffith, Paddy

1987 *Battle Tactics of the Civil War*. Yale University Press, New Haven, Connecticut.

Griffith, Samuel B.

1978 *Mao-Tse-Tung on Guerrilla Warfare*. Double Day, Garden City, New York.

Gwinn, Samuel

1833–1835 Revolutionary War Pension Application S17992. National Archives and Records Administration, Washington, DC.

Haecker, Charles M.

2001 A Well-Laid Trap: Artifact Pattern Analysis of an Apache Ambush Site, Apache Pass, Arizona. Paper presented at the 34th Conference on Historical and Underwater Archaeology, Long Beach, California.

2002 *Archeological Sample Survey of the Florida Mountains, Luna County, New Mexico*. National Park Service, Intermountain Support Office, Santa Fe, New Mexico.

Haecker, Charles M., Elizabeth A. Oster, Angelica Medrano Enríquez, and Michael L. Elliot

2007 Indian Resistance in New Spain: The 1541 AD Battlefield of Peñol de Nochistlán, an Exemplar of Indigenous Resistance. In *Fields of Conflict: Battlefield Archaeology from the Roman Empire to the Korean War: 1. Searching for War in the Ancient and Early Modern World*, edited by Douglas D. Scott, Lawrence Babits, and Charles Haecker, pp. 174–192. Praeger Security International, Westport, Connecticut.

Hafen, Reuben LeRoy, and Ann W. Hafen (editors)

1958 *The Utah Expedition, 1857–1858: A Documentary Account of the United States Military Movement under Colonel Albert Sidney Johnston, and the Resistance by Brigham Young and the Mormon Nauvoo Legion*. A. H. Clark, Spokane, Washington.

Hall, Loretta

2017 Iroquois Confederacy. World Culture Encyclopedia. http://www.everyculture.com/multi/Ha-La/Iroquois-Confederacy.html, accessed May 15, 2017.

Halleck, Henry W.
1862 *Elements of Military Art and Science; or, Course of Instruction in Strategy, Fortification, Tactics of Battles*. D. Appleton, New York.

Halushchak, M.
2013 *Sotnyk UPA Myroslav Symchych: "Skil'ky vstyh—stil'ky vbuv"* [UPA Captain Myroslav Symchych: "How much time—that many killed"]. *Ukrain-s'ka Pravda*, October 14, 2013. http://www.istpravda.com.ua/articles/2013/10/14/137680/, accessed April 15, 2018.

Hanmer, James Williams
1997 The Making of an American Myth: The Hatfields and McCoys. Master's thesis, Department of History, San Jose State University. University Microfilms, Ann Arbor, Michigan.

Hardesty, Donald L.
1997 *The Archaeology of the Donner Party*. University of Nevada Press, Reno.

Hardesty, Hiram H.
1883 *Historical and Genealogical Encyclopedia of West Virginia*. Privately published, New York.

Hardesty, Ryan
2015 Preface. *The Dramatic Story of a Mountain Feud*, by John R. Spears, edited by Ryan Hardesty, unpaginated. Appalachian Heritage Series Vol. 5. Homespun Press, CreateSpace, North Charleston, South Carolina.

Hardin, Albert N., Jr.
1987 *The American Bayonet, 1776–1964*. Privately published, Pennsauken, New Jersey.

Harris, Michael C.
2014 *Brandywine: A Military History of the Battle That Lost Philadelphia but Saved America, September 11, 1777*. Savas Beatie, El Dorado Hills, California.

Harvard University
2015 Andrew W. Mellon Foundation Seminar on Violence and Non-Violence. Asymmetric Warfare: A Symposium. Panel 1: Strategy, Law, and Narrative. Harvard University, April 10, 2015. https://www.youtube.com/watch?v=GaiqpVsF4rM, accessed October 28, 2015.

Hatfield, Coleman C., and Robert Y. Spence
2003 *The Tale of the Devil: The Biography of Devil Anse Hatfield from Original Manuscripts by Grandson Coleman A. Hatfield*. Quarrier Press, Charleston, West Virginia.

Hawkins, Bruce R., and David B. Madsen
1990 *Excavation of the Donner-Reed Wagons: Historical Archaeology along the Hastings Cutoff*. University of Utah Press, Salt Lake City.

Haynes, Dennis E.
2006 *A Thrilling Narrative: The Memoir of a Southern Unionist*. University of Arkansas Press, Fayetteville.

Hegmon, Michelle
2002 Concepts of Community in Archaeological Research. In *Seeking the*

Center Place: Archaeology and Ancient Communities in the Mesa Verde Region, edited by Mark D. Varien and Richard H. Wilshusen, pp. 263–279. University of Utah Press, Salt Lake City.

Heilbrunn, Otto

1962 *Partisan Warfare*. Praeger Books, New York.

Hill, Sam

1888 Report from the Adjutant General of Kentucky. Kentucky National Guard History, Kentucky Documents Collection, Kentucky Department of Library and Archives, Frankfort.

Hill, William

1921 *Col. William Hill's Memoirs of the Revolution*. Edited by A. S. Salley Jr. Historical Commission of South Carolina, State, Columbia, South Carolina.

Hilliard, Jerry E.

2004 *An Antebellum Ozark Community and the Civil War: The Archaeology of the Second Mount Comfort Church (3WA880), Washington County, Arkansas (1840–ca. 1865)*. Arkansas Archeological Survey, Fayetteville.

Hogg, Ian

1982 *The Cartridge Guide*. Stackpole Books, Harrisburg, Pennsylvania.

Houlding, J. A.

1981 *Fit for Service: The Training of the British Army 1715–1795*. Clarendon Press, Oxford.

Howard, James H.

1981 *Shawnee: The Ceremonialism of a Native American Tribe*. Ohio University Press, Athens.

Hoyt, Epaphras

1811 *Practical Instructions for Military Officers: Comprehending A Concise System of Military Geometry, Field Fortification and Tactics of Riflemen and Light Infantry. Also The Scheme for Forming a Corps of a Partisan, and Carrying on the Petite Guerre, by Roger Stevenson, Esq. Revised, Corrected, and Enlarged*. John Denio, Greenville, Massachusetts.

Hughes, Michael A.

2000 Wartime Gristmill Destruction in Northwest Arkansas and Military Farm Colonies. In *Civil War Arkansas: Beyond Battles and Leaders*, edited by Anne Bailey and Daniel E. Sutherland, pp. 31–46. University of Arkansas Press, Fayetteville.

Hungarian Ring

ca. 2015 Austro-Hungarian Machine Gun—Schwarzlose. Electronic document, http://www.hungariae.com, accessed August 29, 2018.

Hunt, George T.

1960 *Wars of the Iroquois: A Study in Intertribal Relations*. University of Wisconsin Press, Madison.

Ilnytzkyj, Roman (editor)

1956 Einsatzkommando Order against the Bandera Movement, dated 25 November 1941. In *Deutschland und die Ukraine, 1934–45*, Vol. 2, pp. 338–339. Oesteuropa-Institut, Munich, Germany.

Isbell, William H.

2000 What We Should Be Studying: The "Imagined Community" and the "Natural Community." In *The Archaeology of Communities: A New World Perspective*, edited by Marcello A. Canuto and Jason Yaeger, pp. 243–266. Routledge, London.

Isley, W. H.

1909 The Sharps Rifle Episode in Kansas History. *American Historical Review* 12(3):546–566.

Jennings, Francis

1984 *The Ambiguous Iroquois Empire*. W. W. Norton, New York.

Joes, Anthony James

1996 *Guerrilla Warfare: A Historical, Bibliographical, and Bibliographical Sourcebook*. Greenwood Press, Westport, Connecticut.

2000 *America and Guerrilla Warfare*. University of Kentucky Press, Lexington.

Johnson, David M., Chris Adams, Charles Hawk, and Skip Keith Miller

2009 *Final Report on the Battle of Cieneguilla, A Jicarilla Apache Victory over the U.S. Dragoons, March 30, 1854*. United States Forest Service, Southwestern Region, Albuquerque, New Mexico.

Johnson, Lewis F.

1916 *Famous Kentucky Tragedies and Trials*. Baldwin Law Book, Louisville, Kentucky.

Johnson, Ludwell H.

1986 "The Horrible Butcheries of West Virginia": Dan Cunningham on the Hatfield-McCoy Feud. *West Virginia History* 46:25–43.

Johnson, Nolan Lee

2006 What Remains? Historic Salvage at the Beaver Creek Trail Crossing Site (25SW49). Master's thesis, Department of Anthropology and Geography, University of Nebraska, Lincoln.

Jones, Virgil Carrington

1948 *The Hatfields and the McCoys*. University of North Carolina Press, Chapel Hill.

Jörgensen, Christer, Michael F. Pavkovic, Rob S. Rice, Frederick C. Schneid, and Chris L. Scott

2005 *Fighting Techniques of the Early Modern World AD 1500–AD 1763: Equipment, Combat Skills, and Tactics*. Thomas Dunn Books, St. Martin's Press, New York.

Keegan, John

1977 *The Face of Battle: A Study of Agincourt, Waterloo & The Somme*. Vintage Books, New York.

Keeley, Lawrence

1996 *War before Civilization: The Myth of the Peaceful Savage*. Oxford University Press, New York.

Kehoe, Alice B.
2006 *North American Indians: A Comprehensive Account.* 3rd ed. Prentice Hall, Upper Saddle River, New Jersey.

Keithly, David
2001 Poor, Nasty and Brutish: Guerrilla Operations in America's First Civil War. *Civil Wars* 4(3):35–69.

Kellogg, Louise Phelps
1916 *Frontier Advance on the Upper Ohio, 1778–1779.* Draper Series Vol. 4. State Historical Society of Wisconsin, Madison.

Kelly, James
1833–1835 Revolutionary War Pension Application S1544. National Archives and Records Administration, Washington, DC.

Kenny, Kathleen M., and John G. Crock
2010 *Hubbardton Battlefield State Historic Site (VT-RU-40): Historical Survey, Property Survey, and ABPP Documentation Project (GA-2255-08-028), Hubbardton, Rutland County, Vermont.* Consulting Archaeology Program, CAP Report #570. Report submitted to National Park Service American Battlefield Protection Program and Vermont Division for Historic Preservation, Montpelier.

King, Dean
2013 *The Feud: The Hatfields and McCoys: The True Story.* Little, Brown, New York.

Knöttel, Herbert, Frederick P. Todd, and John R. Elting
1974 Hesse-Cassel Field Jäger Corps, 1776–1783. In *Military Uniforms in America: The Era of the American Revolution 1755–1795*, edited by John R. Elting, pp. 64–65. Presidio Press, San Rafael, California.

Kolb, Michael J., and James E. Snead
1997 It's a Small World After All: Comparative Analyses of Community Organization in Archaeology. *American Antiquity* 62(4):609–628.

Kosyk, Woodymyr
1993 *The Third Reich and Ukraine.* Studies in Modern European History Vol. 8. Peter Lang, New York.

Kozikowski, Kara E.
2017 Hometown Heroes and Villains. Civil War Preservation Trust. www.civilwar.org/learn/articles/guerrilla-warfare, accessed May 15, 2017.

Kruijt, Dirk
2008 *Guerrillas: War and Peace in Central America.* Zed Books, London.

Lane, Kris, and Timothy F. Johnson (editors)
2007 *The Defense of Western Conquests of Bernardo de Vargas Machuca.* Pennsylvania State University Press, University Park.

Laqueur, Walter
1976 *Guerrilla: A Historical and Critical Study.* Little, Brown, Boston.

Laumbach, Karl W.
2001 *Hembrillo, an Apache Battlefield of the Victorio War: The Archaeology*

of the Hembrillo Battlefield. Human Systems Research. Report prepared for White Sands Missile Range, New Mexico.

La Vere, David

1998 *The Caddo Chiefdoms: Caddo Economics and Politics, 700–1835*. University of Nebraska Press, Lincoln.

LeBlanc, Steven A., and Glen E. Rice

2001 Southwestern Warfare: The Value of Case Studies. In *Deadly Landscapes: Case Studies in Prehistoric Southwestern Warfare*, edited by Glen E. Rice and Steven A. LeBlanc, pp. 1–18. University of Utah Press, Salt Lake City.

Lee, Henry

1998 [1812] *The Revolutionary War Memoirs of General Henry Lee*. Edited by Robert E. Lee. De Capo Press, New York. Originally published as *Memoirs of the War in the Southern Department of the United States*.

Lee, Wayne E.

2004 Fortify, Fight, or Flee: Tuscarora and Cherokee Defensive Warfare and Military Cultural Adaptation. *Journal of Military History* 68:713–770.

Legg, James B., Steven D. Smith, and Tamara S. Wilson

2005 *Understanding Camden: The Revolutionary War Battle of Camden as Revealed through Historical, Archaeological and Private Collections Analysis*. South Carolina Institute of Archaeology and Anthropology. Report presented to the Palmetto Conservation Foundation and the American Battlefield Protection Program.

Lieber, Francis

1898 *Instructions for the Government of Armies of the United States in the Field*. Government Printing Office, Washington, DC.

Linder, Suzanne Cameron, and Emily Linder Johnson

2000 *A River in Time: The Yadkin-Pee Dee River System*. Palmetto Conservation Foundation, Columbia, South Carolina.

Lipe, William

1970 Anasazi Communities in the Red Rock Plateau, Southwestern Utah. In *Reconstructing Prehistoric Pueblo Societies*, edited by William A. Longacre, pp. 84–139. University of New Mexico Press, Albuquerque.

1992 *The San Canyon Archaeological Project: A Progress Report*. Crow Canyon Archaeological Center Occasional Paper No. 2. Crow Canyon Archaeological Center, Cortez, Colorado.

Litteer, Loren K.

1998 *The Life and Times of John Brown*. Champion Publishing, Baldwin, Kansas.

2002 *Bleeding Kansas: The Border War in Douglas and Surrounding Counties*. Champion, Baldwin, Kansas.

Logan, Samuel

2009 *This Is for the Mara Salvatrucha: Inside the MS-13, America's Most Violent Gang*. Hyperion, New York.

Ludwig, Larry

2016 *Archaeological Findings of the Battle of Apache Pass, Fort Bowie National*

Historic Site. National Park Service, Southeast Arizona Group Report NPS/FOBO/SRR—2015/001. Fort Collins, Colorado.

McBride, Kim A., and Tiffany Little Shipp

2015 *Archaeological Investigations at the Randall and Sarah (Sally) McCoy Site, 15PI228, Pike County, Kentucky*. Kentucky Archaeological Survey Report No. 272, Lexington.

McBride, W. Stephen, and Kim A. McBride

1998 Archaeological Investigations of Fort Arbuckle. *Journal of the Greenbrier Historical Society* 6(6):15–45.

2006 Archaeological Investigations of Fort Donnally. *Journal of the Greenbrier Historical Society* 8(2):21–36.

2014 *Frontier Defense: Colonizing Contested Areas in the Greenbrier Valley of West Virginia*. West Virginia Humanities Council, Charleston and Warner, Nicholasville, Kentucky.

McChristian, Douglas C.

2005 *Fort Bowie, Arizona, Combat Post of the Southwest, 1858–1894*. University of Oklahoma Press, Norman.

McCoy, Ron

2015 *Hatfields and McCoys Reunion*. Ferguson Creek, Greensboro, North Carolina.

McCoy, Trudy Williams

1976 *The McCoys: Their Story as Told to the Author by Eye Witnesses and Descendants*. Preservation Council Press, Pikeville, Kentucky.

McCrady, Edward

1902 *The History of South Carolina in the Revolution 1780–1783*. 2 vols. Macmillan and Sons, New York.

McDonald, Robert M. S.

2004 *Thomas Jefferson's Military Academy*. University of Virginia Press, Charlottesville.

McGuire, Thomas J.

2006 *The Philadelphia Campaign: I. Brandywine and the Fall of Philadelphia*. Stackpole Books, Mechanicsburg, Pennsylvania.

2007 *The Philadelphia Campaign: II. Germantown and the Roads to Valley Forge*. Stackpole Books, Gettysburg, Pennsylvania.

Mackey, Robert R.

2004 *The Uncivil War: Irregular Warfare in the Upper South, 1861–1865*. University of Oklahoma, Norman.

McKee, William

1778 Letter to General Edward Hand, June 21. Draper Manuscripts 3NN210–213. State Historical Society of Wisconsin, Madison. (Reprinted in Kellogg 1916:98–99.)

McKnight, Brian D.

2006 *Contested Borderland: The Civil War in Appalachian Kentucky and Virginia*. University Press of Kentucky, Lexington.

2009 "Time by the Forelock": Champ Ferguson and the Borderland Style of Warfare. In *Sister States Enemy States: The Civil War in Kentucky and Tennessee*, edited by Kent T. Dollar, Larry H. Whiteaker, and W. Calvin Dickinson, pp.140–167. University Press of Kentucky, Lexington.

McNeilly, Donald P.

2000 *The Old South Frontier: Cotton Plantations and the Formation of Arkansas Society, 1819–1861*. University of Arkansas Press, Fayetteville.

McPheeters, William

2002 *"I Acted from Principle": The Civil War Diary of Dr. William M. McPheeters, Confederate Surgeon in the Trans-Mississippi*. University of Arkansas Press, Fayetteville.

McWhiney, Grady, and Perry D. Jamieson

1984 *Attack and Die: Civil War Military Tactics and the Southern Heritage*. University of Alabama Press, Tuscaloosa.

Magruder, John B.

1893 Letter to Senator Robert W. Johnson. In *The War of the Rebellion: A Compilation of the Official Records of the Union and Confederate Armies*. Series 1, vol. 41, chap. 53, pt. 4, edited by George B. Davis, pp. 1029–1031. Government Printing Office, Washington, DC.

Mahan, Dennis H.

1836 *A Complete Treatise on Field Fortification: With the General Outlines of the Principles Regulating the Arrangement, the Attack, and the Defense of Permanent Works*. Wiley & Long, New York.

Mahon, John K.

1958 Anglo-American Methods of Indian Warfare, 1676–1794. *Mississippi Valley Historical Review* 45:254–275.

1960 The Journal of A.B. Meek and the Second Seminole War, 1836. *Florida Historical Quarterly* 38(4):302–318.

1985 *History of the Second Seminole War*. Rev ed. University of Florida Press. Gainesville.

Malone, Patrick M.

1991 *The Skulking Way of War: Technology and Tactics among the New England Indians*. Madison Books, Lanham, Maryland.

Mandel, Rolf D.

2006 Late Quaternary and Modern Environments in Kansas. In *Kansas Archaeology*, edited by Robert J. Hoard and William E. Banks, pp. 10–27. University Press of Kansas, Lawrence.

Mandzy, Adrian

2007 Tatars, Cossacks, and the Polish Army: The Battle of Zboriv. In *Fields of Conflict: Battlefield Archaeology from the Roman Empire to the Korean War: 1. Searching for War in the Ancient and Early Modern World*, edited by Douglas D. Scott, Lawrence Babits, and Charles Haecker, pp. 193–207. Praeger Security International, Westport, Connecticut.

2012 Using Munitions and Unit Frontage: New Evidence about the Russian

Main Battle Line at Poltava (1709). *Fasciculi Archaeologiae Historicae* 25:67–76.

2016 Oral Testimony and Battlefield Archaeology: A Multi-Disciplinary Study of a Ukrainian Ambush from 1945. In *Papers from the 2014 Fields of Conflict Conference and Preservation Workshop*, edited by Steven D. Smith, pp. 199–202. South Carolina Institute of Archaeology and Anthropology, Columbia.

Matson, Daniel S., and Albert H. Schroeder (editors and translators)

1957 Cordero's Description of the Apaches—1796. *New Mexico Historical Review* 32(4):335–356.

Mattes, Merrill J.

1987 *The Great Platte River Road: The Covered Wagon Mainline via Fort Kearny to Fort Laramie*. Bison Books, University of Nebraska Press, Lincoln.

Medrano, Angélica María Enriquez

2012 *Arqueología Del Conflicto: La Guerra del Mixtón (1541–1542) vista a través del Peñol de Nochistlán*. Taberna Libraria Editores, Zacatecas, Mexico.

Metz, Steven, and Douglas V. Johnson II

2001 *Asymmetry and U.S. Military Strategy: Definition, Background, and Strategic Concepts*. Strategic Studies Institute, United States Army War College, Carlisle, Pennsylvania. http://www.strategicstudiesinstitute.army.mil/pdffiles/PUB223.pdf, accessed February 18, 2016.

Miller, Jeremy, Álvaro Ibarra, and M. Scott Harris

2014 *Combating Asymmetrical Operations: Understanding the Roman Occupation of Dacia*. Paper presented at the 8th Fields of Conflict Biennial Conference, Columbia, South Carolina.

Mills, Robert

1980 [1825] *Mills' Atlas of the State of South Carolina*. Reprinted. Southern Historical Press, Greenville, South Carolina.

Missal, John, and Mary Lou Missal

2005 *This Miserable Pride of a Soldier*. University of Tampa Press, Tampa, Florida.

Monaghan, Jay

1984 *Civil War on the Western Border, 1854–1865*. Bison Books, University of Nebraska, Lincoln.

Moneyhon, Carl H.

1993 Disloyalty and Class Consciousness in Southwestern Arkansas, 1862–1865. *Arkansas Historical Quarterly* 52(3):223–243.

2000 Disloyalty and Class Consciousness in Southwestern Arkansas, 1862–1865. In *Civil War Arkansas: Beyond Battles and Leaders*, edited by Anne J. Bailey and Daniel E. Sutherland, pp. 117–142. University of Arkansas Press, Fayetteville.

Montell, William Lynwood

1986 *Killings: Folk Justice in the Upper South*. University Press of Kentucky, Lexington.

Moore, Maurice A.
1859 *The Life of General Edward Lacey*. Evins, Spartanburg, South Carolina.

Moorehead, Max L.
1975 *The Presidio, Bastion of the Spanish Borderland*. University of Oklahoma Press, Norman.

Morton, Oren
1916 *A History of Monroe County, West Virginia*. Reubush-Elkins, Dayton, Virginia.

Mosby, John Singleton
2006 *Mosby's Memoirs: The Memoirs of Colonel John Singleton Mosby*. Barnes and Noble, New York.

Mountcastle, Clay
2009 *Punitive War: Confederate Guerrillas and Union Reprisals*. University Press of Kansas, Lawrence.

Mounts, Ellison [Ellison Hatfield, pseud.]
1888 Full testimony Given by Ellison Mounts and Recorded by J. Lee Ferguson. Governor Simon B. Buckner, Correspondence, Record Group 1189, Acc #1984-194, Box 1, Folder 16. Kentucky Department for Libraries and Archives, Frankfort.

Muenchhausen, Friedrich von
1974 *At General Howe's Side, 1776–1778: The Diary of General William Howe's Aide de camp, Captain Friedrich von Muenchhausen*. Translated by Ernst Kipping and annotated by Samuel S. Smith. Philip Freneau Press, Monmouth Beach, New Jersey.

Munnerlyn, L.
1833–1837 Revolutionary War Pension Application S18136. National Archives and Records Administration, Washington, DC.

Muñoz, Col.
1541 Document XI, Carta de don Antonio de Mendoza al Obispo de Mexico. De junto al Coyna, Bancroft Library, University of California, Berkeley.

Murdock, G. P., C. S. Ford, A. E. Hudson, R. Kennedy, L. W. Simmons, and J. W. M. Whiting
1945 *Outline of Cultural Materials*. Yale Anthropological Series II. New Haven, Connecticut.

Nauvoo Tourism Office
2017 Nauvoo's Mormon Period (1839–1846). Electronic document, https://www.beautifulnauvoo.com/nauvoo-during-the-mormon-period-(1839-1846).html, accessed May 15, 2017.

Neumann, George C.
1967 *The History of Weapons of the American Revolution*. Bonanza Books, New York.

Noe, Kenneth W.
1997 Exterminating Savages: The Union Army and Mountain Guerillas in

Southern West Virginia, 1861–1862. In *The Civil War in Appalachia: Collected Essays*, edited by Kenneth Noe, pp. 104–130. University of Tennessee Press, Knoxville.

Noe, Kenneth W. (editor)
1997 *The Civil War in Appalachia: Collected Essays*. University of Tennessee Press, Knoxville.

Noe, Kenneth W., and Shannon H. Wilson
1997 Introduction. In *The Civil War in Appalachia: Collected Essays*, edited by Kenneth W. Noe, pp. xi–xxxiv. University of Tennessee Press, Knoxville.

Nowak, Marian
2003 Suplement do lusek austriakich. *Odkrywca* 3(50):46–47.

O'Bryan, Tony
2017 Bushwhackers. *Civil War on the Western Border: The Missouri-Kansas Conflict, 1854–1865*. Kansas City Public Library, Kansas City. http://www.civilwaronthewesternborder.org/encyclopedia/bushwhackers, accessed June 7, 2017.

Ogle, George O.
1902 *Standard Atlas of Douglas County, Kansas*. George O. Ogle, Chicago.
1921 *Standard Atlas of Douglas County, Kansas*. George O. Ogle, Chicago.

Onyshchuk, Iaroslav
2013 Doslidzhennia pokhovannia bijtsiv Karpats'koi Ukrainy na Verets'komy perevali v 2011 rotsi [Research of the burial places of Carpathian Ukraine's soldiers on Veretsky Mountain Pass in 2011]. *Tsytadelia* [Citadel] 9:46–50.

Otterbein, Keith F.
2000 Five Feuds: An Analysis of Homicides in Eastern Kentucky in the Late Nineteenth Century. *American Anthropologist* 102(2):231–243.
2004 *How War Began*. Texas A& M University Press, College Station.
2009 *The Anthropology of War*. Waveland Press, Long Grove, Illinois.

Ovsienko, V. V.
2002 Z vohniu ta v polum'ia: Interv'iu z povstantsem Myroslavom Symchychem [From fire and into flame: Interview with Myroslav Vasyl'ovych Symchych]. *Suchasnist'* 2(490):132–148; 3(491):57–80. http://archive.khpg.org/index.php?id=1121371877, accessed October 21, 2015.

Pancake, John S.
1985 *This Destructive War: The British Campaign in the Carolinas, 1780–1782*. University of Alabama Press, Tuscaloosa.

Patryliak, Ivan
2012 *"Vstan' i borya'! Slukhaj i vir . . .": Ukrains'ke Nationalistychne Pidpillia ta Povstans'kyj Rukh (1939–1960)* ["Stand and fight! Listen and believe . . .": Ukrainian nationalistic and insurgent movement (1939–1960)]. Tsentr doslidzhen' vyzvol'noho rukhu, L'viv, Ukraine.

Patton, Samuel
1830 Pension Application S31,901. Record Group M804, National Archives and Records Administration, Washington, DC.

Pearce, John Ed
1994 *Days of Darkness: The Feuds of Eastern Kentucky*. University of Lexington Press, Lexington, Kentucky.

Peebles, John
1998 *John Peebles' American War: The Diary of a Scottish Grenadier, 1776–1782*. Edited by Ira D. Gruber. Stackpole Books, Mechanicsburg, Pennsylvania.

Pennsylvania Gazette [Philadelphia]
1777 Extract of a September 9 letter from headquarters. 10 September. Philadelphia, Pennsylvania.

Peterson, Harold L.
1968 *The Book of the Continental Soldier*. Stackpole, Harrisburg, Pennsylvania.

Pfefferkorn, Ignaz
1989 *Sonora: A Description of the Province*. Edited and translated by Theodore E. Treutlein. University of Arizona Press, Tucson.

Piecuch, Jim
2008 *Three Peoples, One King: Loyalists, Indians, and Slaves in the Revolutionary South, 1775–1782*. University of South Carolina Press, Columbia.

Poll, Richard D., and Ralph W. Hansen
1961 "Buchanan's Blunder": The Utah War, 1857–1858. *Military Affairs* 25(3): 121–131.

Porter, Charles W.
1998 *In the Devil's Dominions: A Union Soldier's Adventures in "Bushwhacker Country."* Bushwhacker Museum, Nevada, Missouri.

Posivnych, Mykola
2010 *Voenno-politychna diial'nist' OUN v 1929–1939* [Military-political actions of the OUN in 1929–1939]. National Academy of Sciences of Ukraine, L'viv.

Powell, John T.
1996 German Military Buttons of the Period of the American War for Independence, 1776–1783. *Military Collector and Historian* 48(3):121–123.

Powell, Philip Wayne
1952 *Soldiers, Indians and Silver: The Northward Advance of New Spain, 1550–1600*. University of California Press, Berkeley.

Prescott, William H.
1992 [1843] *History of the Conquest of Mexico*. Folio Society, London.

Preucel, Robert W.
2000 Making Pueblo Communities: Architectural Discourse at Kotyiti, New

Mexico. In *The Archaeology of Communities: A New World Perspective*, edited by Marcello A. Canuto and Jason Yaeger, pp. 58–77. Routledge, London.

Price, Sterling

1891 Special Orders No. 50. In *The War of the Rebellion: A Compilation of the Official Records of the Union and Confederate Armies*. Series 1, vol. 34, chap. 46, pt. 2, p. 1001. Government Printing Office, Washington, DC.

Purdy, Gilbert

1777–1778 Memorandum of Lt. Gilbert Purdy. Typed copy by Albert J. Wright. Original at National Archives of Canada (Z6620/C21/1975/U2) Ref.: PAC, M623 (B14). On file at Chester County Historical Society, West Chester, Pennsylvania.

Rada, James, Jr.

2013 *Kidnapping the Generals: The South's Most-Daring Raid against the Union Army*. Legacy, Charleston, South Carolina.

Ramage, James A.

2009 *Gray Ghost: The Life of Col. John Singleton Mosby*. University Press of Kentucky, Lexington.

Rankin, Hugh F.

1973 *Francis Marion: The Swamp Fox*. Thomas Y. Crowell, New York.

Rawson, Glenn, and Dennis Lyman (editors)

2014 *The Mormon Wars Early Persecutions, Hawn's Mill, Nauvoo War, Johnston's Army, War on Polygamy*. Deseret Book, Littleton, Colorado.

Redpath, James

1860 *The Public Life of Capt. John Brown, with an Autobiography of his Childhood and Youth*. Harper and Eldridge, Boston.

Restall, Matthew

2004 *Seven Myths of the Spanish Conquest*. Oxford University Press, New York.

Retzer, Henry (translator), and Donald Londahl-Smidt (annotations)

1995 Letters of Friedrich A. J. von Wagenheim, Lieutentant and Staff Captaini, Hesse Kassel Fieldjager Corps. *Journal of the Johannes Schwalm Association* 6(2):1–25.

1998 The Philadelphia Campaign, 1777–1778: Letters and Reports from the von Jungkenn Papers. *Journal of the Johannes Schwalm Association* 5(3):1–7.

Rice, Otis K.

1970 *The Allegheny Frontier: West Virginia Beginnings, 1730–1830*. University Press of Kentucky, Lexington.

1982 *The Hatfields and McCoys*. University Press of Kentucky, Lexington.

1986 *History of Greenbrier County*. McClain Printing, Parsons, West Virginia.

Robbins, James S.

2006 *Last in Their Class: Custer, Pickett and the Goats of West Point*. Encounter Books, New York.

Robertson, Archibald

1930 *Archibald Robertson, Lieutenant-General Royal Engineers: His Diaries and Sketches in America, 1762–1780*. Edited by Harry Miller Lyndberg. New York Public Library, New York.

Rodgers, Sarah

ca. 1900 Viney Family by Mrs. Sallie Rodgers, as told to her by her mother Sallie Rodgers. Manuscript on file, Museum of Anthropology, University of Kentucky, Lexington.

Ross, Steven

1979 *From Flintlock to Rifle: Infantry Tactics, 1740–1866*. Fairleigh Dickinson University Press, Rutherford, New Jersey.

Royall, Anne

1991 [1826] Sketches of Life. Reprinted extract. *Journal of the Greenbrier Historical Society* 4(1):52–69.

Salley, A. S., Jr. (editor)

1900 Letter of Thomas Post [Potts] to the Council of Safety in Charles Town. Papers of the First Council of Safety of the Revolutionary Party in South Carolina, June–November 1775. *South Carolina Historical and Genealogical Magazine* 1(2):128–129.

Sassaman, Kenneth E.

2001 Hunter-Gatherers and Traditions of Resistance. In *The Archaeology of Traditions: Agency and History before and after Columbus*, edited by Timothy R. Pauketat, pp. 218–236. University Press of Florida, Gainesville.

Saunders, Bruce D.

1989 Lessons Learned: Apache Campaign Staff Ride. *Military Intelligence*. October–December:43–44.

Schram, Stuart R. (translator)

1966 *Mao Tse-Tung Basic Tactics*. Frederick A. Praeger, New York.

Schroeder, Albert E.

1983 Querechos, Vaqueros, Cocoyes, and Apaches. In *Collected Papers in Honor of Charlie Steen Jr.*, edited by Nancy Fox, pp. 159–166. Archaeological Society of New Mexico Papers No. 8.

Schroeder, Susan (editor)

1998 *Native Resistance and the Pax Colonial in New Spain*. University of Nebraska Press, Lincoln.

Scoggins, Michael C.

2005 *The Day It Rained Militia: Huck's Defeat and the Revolution in the South Carolina Backcountry, May–July 1780*. History Press, Charleston, South Carolina.

Scoggins, Michael C., Steven D. Smith, and Tamara Wilson

2011 *Defining the Williamson's Plantation: Huck's Defeat Battlefield*. South Carolina Institute of Archaeology and Anthropology, Columbia.

Scott, Douglas D.

2007 *An Episode of the Bleeding Kansas Era: Archaeological Investigations of the Black Jack Battlefield and Ivan Boyd Prairie, Douglas County, Kan-*

sas. Connor Consulting. Report prepared for the Black Jack Battlefield Trust. Copy on file with Office of the State Archeologist, Kansas State Historical Society, Topeka.

Scott, Douglas D. (editor)

1991 *Papers on Little Bighorn Battlefield Archaeology: The Equipment Dump, Marker 7, and the Reno Crossing*. Reprints in Anthropology No. 42. J and L Reprint, Lincoln, Nebraska.

Scott, Douglas D., and Richard A. Fox Jr.

1987 *Archeological Insights into the Custer Battle: A Preliminary Assessment*. University of Oklahoma Press, Norman.

Scott, Douglas D., Richard A. Fox Jr., Melissa A. Connor, and Dick Harmon

1989 *Archaeological Perspectives on the Battle of the Little Bighorn*. University of Oklahoma Press, Norman.

Scott, Douglas D., and Lucien Haag

2009 "Listen to the Minié Balls": Identifying Firearms in Battlefield Archaeology. In *Fields of Conflict: Battlefield Archaeology from Imperial Rome to Korea*, edited by Douglas D. Scott, Lawrence Babits, and Charles Haecker, pp. 102–120. Potomac Books, Washington, DC.

Scott, Douglas D., Charles Haecker, and Lawrence Babits (editors)

2007 *Fields of Conflict: Battlefield Archaeology from the Roman Empire to the Korean War*. 2 vols. Praeger Security International, New York. (Republished as a single volume, Potomac Books, Washington DC, 2009.)

Scott, Douglas D., Thomas D. Theissen, and Steve J. Dasovich

2014 *A "Desperate and Bloody" Fight: The Battle of Moore's Mill, Callaway County, Missouri, July 28, 1862*. Missouri's Civil War Heritage Foundation. Report prepared for the American Battlefield Protection Program, National Park Service, Washington, DC.

Secoy, Frank R.

1953 *Changing Military Patterns on the Great Plains*. Monographs of the American Ethnological Society Vol. 21. J. J. Augustin, New York.

Selig, Robert A.

1994 Light Infantry Lessons from America? Johann Ewald's Experiences in the American War as Depicted in his *Abhandlung über den Kleinen Krieg* (1785). *Studies in Eighteenth Century Culture* 23:111–129.

Selig, Robert A., Christopher T. Espenshade, Wade P. Catts, Elisabeth Lavigne, and James Montney

2016 *Cultural Resources Survey of the Bennington Battlefield. Walloomac, New York*. Commonwealth Heritage Group. Prepared for the New York State Office of Parks, Recreation and Historic Preservation, Albany.

Selig, Robert A., and David Curtis Skaggs (editors and translators)

1991 *Treatise on Partisan Warfare*, by Johann von Ewald. Contributions in Military Studies No. 116. Greenwood Press, New York.

Sesser, David

2012 Skirmish at McGraw's Mill. *Encyclopedia of Arkansas History & Culture*. Butler Center for Arkansas Studies at the Central Arkansas Library System, Little Rock. http://www.encyclopediaofarkansas.net/encyclopedia/entry-detail.aspx?search=1&entryID=6691, accessed August 8, 2018.

Shchehliuk, Muron.

2014 *Strilets'ka zbroia Ukrains'kykh povstantsiv* [Firearms of the Ukrainian Partisans]. Edited by Ye Shtendera and Peter Potichni. Spolom, L'viv, Ukraine.

Simms, William Gilmore

1845 *The Life of Francis Marion*. Henry G. Langley, New York.

Sivilich, D. M.

1996 Analyzing Musket Balls to Interpret a Revolutionary War Site. *Journal of the Society for Historical Archaeology* 30(2):101–109.

2008 *Cultural Resources Report—Electronic Phase I Archaeological Survey Conducted at: W. L. Gore & Associates, Inc. Properties, Parcels B (11-0.18.00-061), D (11-0.18.00 059), and F (11-0.22.00-0.15), New Castle County, Delaware*. Battlefield Restoration and Archaeological Recovery Organization. Report prepared for W. L. Gore and Associates, Newark, Delaware.

2009 What the Musket Ball Can Tell: Monmouth Battlefield State Park, New Jersey. In *Fields of Conflict: Battlefield Archaeology from the Roman Empire to the Korean War*, edited by Douglas D. Scott, Lawrence Babits, and Charles Haecker, pp. 84–101. Potomac Books, Washington, DC.

Sivilich, Michelle

2012 A Proposed Model to Investigate the Role of Education in the Success of Military Strategy in Florida during the Second Seminole War (1835–1842). *Historical Archaeology* 46(1):62–73.

2014 Measuring the Adaptation of Military Response during the Second Seminole War Florida (1835–1842): KOCOA and The Role of a West Point Military Academy Education. PhD dissertation, Anthropology Department, University of South Florida, Tampa.

Skelton, William B.

2004 West Point and Officer Professionalism, 1817–1877. In *West Point: Two Centuries and Beyond*, edited by Lance Betros, pp. 22–37. McWhiney Foundations Press, Texas A& M University Press, College Station.

Smith, F. Todd

1998 The Political History of the Caddo Indians: 1686–1874. In *The Native History of the Caddo: Their Place in Southeastern Archeology and Ethnohistory*, edited by Timothy K. Perttula and James E. Bruseth. Studies in Archeology No. 30. Texas Archeological Research Laboratory, Austin.

Smith, James

1948 [1812] *A Treatise on the Mode and Manner of Indian War*. J. R. Lyle, Paris, Kentucky. 1948 facsimile ed. Newberry Library, Chicago.

Smith, Samuel S.

1976 *The Battle of the Brandywine*. Philip Freneau Press, Monmouth Beach, New Jersey.

Smith, Steven D.

2010 Archaeological Perspectives on Partisan Communities: Francis Marion at Snow's Island in History, Landscape, and Memory. PhD dissertation, Department of Anthropology, University of South Carolina, Columbia.

Smith, Steven D. (editor)

2016 *Preserving Fields of Conflict: Papers from the 2014 Fields of Conflict Conference and Preservation Workshop*. South Carolina Institute of Archaeology and Anthropology, University of South Carolina, Columbia.

Smith, Steven D., James B. Legg, and Tamara S. Wilson

2007 *The Search for Williamson's Plantation: Huck's Defeat Battlefield*. South Carolina Institute of Archaeology and Anthropology, Columbia.

2009 *The Archaeology of the Camden Battlefield: History, Private Collections, and Field Investigations*. South Carolina Institute of Archaeology and Anthropology, Columbia. Report prepared for the Palmetto Conservation Foundation and Save America's Treasures, National Park Service, Washington, DC.

Smith, Steven D., James B. Legg, Tamara S. Wilson, and Jonathan Leader

2007 *"Obstinate and Strong": The History and Archaeology of the Siege of Fort Motte*. South Carolina Institute of Archaeology and Anthropology, Columbia. Report prepared for the American Battlefield Protection Program, National Park Service, Washington, DC.

Smith, Steven D., Tamara S. Wilson, and James B. Legg

2008 *The Search for Francis Marion: Archaeological Survey of 15 Camps and Battlefields associated with Francis Marion*. South Carolina Institute for Archaeology and Anthropology, Columbia. Report prepared for the Francis Marion Trail Commission, Florence, South Carolina.

Smith, William W.

1836 *Sketch of the Seminole War and Sketches during a Campaign*. Dan J. Dowling, Charleston, South Carolina.

Sodol, Petro R.

1987 *UPA: They Fought Hitler and Stalin*. Committee for the World Convention and Reunion of Soldiers in the Ukrainian Insurgent Army, New York.

Sokhan, P. Y. Shtendera, and V. Ivanenko (editors)

1995 *Publications of the UPA Supreme Command*. Litopys UPA. New series, vol. 1. National Academy of Sciences of Ukraine, Kiev.

Sosin, Jack M.

1967 *The Revolutionary Frontier, 1763–1783*. Holt, Rinehart and Winston, New York.

Soudes, D. B. M.

1909 *Plat Work and Complete Survey of Douglas County, Kansas*. Kenyon, Des Moines, Iowa.

Sparks, George, Patricia Wolf, and Andryce M. Zurick

2015 Destructive Leadership: The Hatfield and McCoy Feud. *American Journal of Business Education* 8(4):307–325.

Spears, John R.
2015 [1888] *The Dramatic Story of a Mountain Feud*. 2015 annotated ed. Edited by Ryan Hardesty. Appalachian Heritage Series Vol. 5. Homespun Press, CreateSpace, North Charleston, South Carolina.

Sprague, John T.
2000 [1848] *The Origin, Progress, and Conclusion of the Florida War*. University of Tampa Press, Tampa, Florida.

Spring, Leverett Wilson
1887 *Kansas, the Prelude to the War for the Union*. Houghton, Mifflin, Boston.

Staton, Willis David
1993 *Hatfields and McCoys: True Romance and Tragedies*. Aegina Press, Huntington, West Virginia.

Stephens, Thomas R.
1998 In Deepest Submission: The Hessian Mercenary Troops in the American Revolution. PhD dissertation, History Department, Texas A&M University, College Station.

Stevens, Jacob
ca. 1840 Interview by the Rev. Dabney Shane. Draper Manuscripts 12CC:133–138. State Historical Society of Wisconsin, Madison.

Stewart, Walter
1777 Walter Stewart to General Gates, Newport [Delaware], September 2. In *Bicentennial Edition of the Spirit of 'Seventy-Six: The Story of the American Revolution as Told by Participants*, edited by Henry Steele Commager and Richard B. Morris, p. 610. Harper and Row, New York.

Stirke, Henry
1961 A British Officer's Revolutionary War Journal, 1776–1778. Edited by S. Sydney Bradford. *Maryland Historical Magazine* 56:150–175.

Stone, Richard
1978 *A Brittle Sword: The Kentucky Militia, 1776–1912*. University Press of Kentucky, Lexington.

Stuart, John
1820 Narrative. Draper Manuscripts 6NN105–112. State Historical Society of Wisconsin, Madison. (Reprinted in Kellogg 1916:70–73.)
1971 [1833] *Memoir of Indian Wars*. 1971 facsimile ed. Collections of the Virginia Historical and Philosophical Society Vol. 1. McClain Printing, Parsons, West Virginia.

Subtelny, Orest
1991 *Ukraine: A History*. University of Toronto Press, Toronto.

Summers, Lewis P.
1929 *Annals of Southwest Virginia*. Privately published, Abingdon, Virginia.

Sutherland, Daniel E.
2009 *Savage Conflict: The Decisive Role of Guerillas in the American Civil War*. University of North Carolina Press, Chapel Hill.
2015 Guerilla Warfare in Virginia during the Civil War. *Encyclopedia Virginia*. Virginia Foundation for the Humanities. https://www.encyclope

diavirginia.org/Guerrilla_Warfare_in_Virginia_during_the_Civil_War, accessed June 7, 2017.

Swepston, Paul

2013 *John Brown's Diary: Journal of Business and Engagements, July 4, 1852–July 13, 1865*. Privately published.

Swope, Michael

1833–1835 Revolutionary War Pension Application R01366. National Archives and Records Administration, Washington, DC.

Tarleton, Banastre

1787 *A History of the Campaigns of 1780 and 1781 in the Southern Provinces of North America*. T. Cadell, London.

Taylor, Hugh Paul

1829 Notes. Draper Manuscripts 32S13–18. State Historical Society of Wisconsin, Madison.

Thiessen, Thomas D., Douglas D. Scott, and Steven J. Dasovich

2008 *"This Work of Fiends": Historical and Archaeological Perspectives on the Confederate Guerrilla Actions at Centralia, Missouri, September 27, 1864*. Report prepared for the Friends of Centralia Battlefield and the Missouri Civil War Heritage Foundation, Lincoln, Nebraska.

Thomas, Brian W.

2001 African-American Tradition and Community in the Antebellum South. In *The Archaeology of Traditions*, edited by Timothy R. Pauketat, pp. 17–33. University Press of Florida, Gainesville.

Thomas, Dean S.

1997 *Round Ball to Rimfire: A History of Civil War Small Arms Ammunition*. Part 1. Thomas, Gettysburg, Pennsylvania.

Thornton, Rod

2007 *Asymmetric Warfare: Threat and Response in the Twenty-First Century*. Polity, Cambridge.

Thrapp, Dan L.

1980 *Victorio and the Mimbres Apaches*. University of Oklahoma Press, Norman.

Thwaites, Reuben G., and Louise P. Kellogg

1905 *Documentary History of Dunmore's War, 1774*. Draper Series Vol. 1. State Historical Society of Wisconsin, Madison.

1908 *The Revolution on the Ohio, 1775–1777*. Draper Series Vol. 2. State Historical Society of Wisconsin, Madison.

1912 *Frontier Defense on the Upper Ohio, 1777–1781*. Draper Series Vol. 3. State Historical Society of Wisconsin, Madison.

Tierney, John J., Jr.

2006 *Chasing Ghosts: Unconventional Warfare in American History*. Potomac Books, Washington, DC.

Townsley, James

1879 Statement of James Townsley. *Republican Citizen* 20 December. Paola, Kansas.

Trident [New York]

1941 Western Ukraine Declares Its Independence. July–August, 5(6):1. New York.

Trigger, Bruce G.

1968 The Determinates of Settlement Patterns. In *Settlement Archaeology*, edited by K. C. Chang, pp. 53–78. National Press Books, Palo Alto, California.

1976 The Disappearance of the St. Lawrence Iroquoians. In *The Children of Aataenstic: A History of the Huron People to 1660*, vol. 2, edited by Bruce Trigger, pp. 214–218. McGill-Queen's University Press, Montreal.

1989 *A History of Archaeological Thought*. Cambridge University Press, New York.

Troiani, Don

2001 *Military Buttons of the American Revolution*. Thomas, Gettysburg, Pennsylvania.

Troiani, Don, Earl J. Coates, and James L. Kochan

1998 *Don Troiani's Soldiers in America 1754–1865*. Stackpole Books, Mechanicsburg, Pennsylvania.

Turney-High, Harry Holbert

1971 *Primitive Warfare: Its Practice and Concepts*. 2nd ed. University of South Carolina Press, Columbia.

Tys-Krokhmaliuk, Yuriy

1972 *UPA Warfare in Ukraine: Strategical, Tactical and Organizational Problems of Ukrainian Resistance in World War II*. Society of Veterans of Ukrainian Insurgence Army, New York.

Uhlendorf, Bernard A., and Edna Vosper (editors)

1935 Letters of Major Baurmeister during the Philadelphia Campaign, 1777–1778. *Pennsylvania Magazine of History and Biography* 59(4):392–419.

1937 *Letters from Major Baurmeister to Colonel von Jungkenn Written during the Philadelphia Campaign, 1777–1778*. Historical Society of Pennsylvania, Philadelphia.

United States Army

2008 *Army Special Operations Forces: Unconventional Warfare*. FM 3-05.130. Headquarters, Department of the Army, Washington, DC. https://fas.org/irp/doddir/army/fm3-05-130.pdf, accessed February 18, 2016.

United States Department of Defense [USDoD]

2010 *Department of Defense Dictionary of Military and Associated Terms*. Joint Publication 1-02. United States Department of Defense, Washington, DC.

2013 *Counterinsurgency*. Joint Publication 3-24. United States Department of Defense, Washington, DC.

United States Government

2017 *Gangs and Crime in America: Transnational Threats from the Mara Salvatrucha MS-13 and 18th Street Gangs, Origins, Relationship to Street

Gangs, Federal Response, El Salvador Initiative, RICO Act. Congressional Research Service. Washington, DC.

Vargas Machuca, Bernardo de

2008 [1599] *The Indian Militia and Description of the Indies*. Edited by Kris Lane. Duke University Press, Durham, North Carolina.

Vauban, Sébastien le Prestre de

1737 *De l'Attaque et de la Defense des Places*. P de Hondt, The Hague.

V'iatrovych, Volodymyr, Ruslan Zabilyj, Ihor Derev'ianyj, and Petro Sodol

2011 *Ukrains'ka Povstans'ka Armia: Istoria neskorenykh* [Ukrainian Partisan Army: History of the undefeated]. Tsentr doslidzhennia vyzvol'noho rykhy, L'viv, Ukraine.

Villard, Oswald Garrison

1910 *John Brown, 1800–1859: A Biography Fifty Years After*. Houghton Mifflin, Boston.

Waghelstein, John D.

1995 Regulars, Irregulars, and Militia: The American Revolution. *Small Wars and Insurgencies* 6(2):133–158.

Wallace, Samuel A., and Yaroslaw Chyz

1931 *Western Ukraine under Polish Yoke: Polonization, Colonization, and "Pacification."* Ukrainian Review, New York.

Waller, Altina L.

1988 *Feud: Hatfields, McCoys, and Social Change in Appalachia, 1860–1900*. University of North Carolina Press, Chapel Hill.

Washington, George

1776 General Orders, June 29. George Washington Papers at the Library of Congress, 1741–1799. Series 3g, Varick Transcripts, letterbook 1, image 297. http://memory.loc.gov/ammem/gwhtml/gwhome.html, accessed April 15, 2012.

1777a General Orders, Head Quarters, Wilmington, August 28, Additional After Orders. George Washington Papers at the Library of Congress, 1741–1799. Series 3g, Varick Transcripts, letterbook 2, image 220. http://memory.loc.gov/ammem/gwhtml/gwhome.html, accessed April 15, 2012.

1777b General Orders, October 6. George Washington Papers at the Library of Congress, 1741–1799. Series 3g, Varick Transcripts, letterbook 1, image 263. http://memory.loc.gov/ammem/gwhtml/gwhome.html, accessed April 15, 2012.

Washington Telegraph [Washington, Arkansas]

1864 To the Citizens and Army of Arkansas. 12 December:2. Washington, Arkansas.

Watts, Dale E.

1998 Plows and Bibles, Rifles and Revolvers: Guns in Kansas Territory. *Kansas History* 21(1):30–45.

Weigley, Russell F.

1970 *The Partisan War: The South Carolina Campaign of 1780–1782*. Tricentennial Booklet No 2. University of South Carolina Press, Columbia.

Weisman, Brent
1989 *Like Beads on a String: A Culture History of the Seminole Indians in Northern Peninsular Florida*. University of Alabama Press, Tuscaloosa.
1999 *Unconquered People: Florida's Seminole and Miccosukee Indians*. University Press of Florida, Gainesville.
2007 Nativism, Resistance, and Ethnogenesis of the Florida Seminole Indian Identity. *Historical Archaeology* 41(4):198–212.

Weller, Jac
1977 Irregular but Effective: Partizan Weapons and Tactics in the American Revolution Southern Theater. In *Military Analysis of The Revolutionary War: An Anthology*, edited by Donald Higginbotham, pp. 131–144. KTO Press, Millwood, New York.

Wheatley, David, and Mark Gillings
2002 *Spatial Technology and Archaeology: The Archaeological Applications of GIS*. Taylor and Francis, New York.

Whinyates, F. A. (editor)
1898 The Services of Lieut.-Colonel Francis Downman, R.A., in France, North America, and the West Indies, between the Years 1758 and 1784. *Minutes of the Proceedings of the Royal Artillery Institution* 25:153–172.

White, Richard
1991 *The Middle Ground: Indians, Empires, and Republics in the Great Lakes Region, 1650–1815*. Cambridge University Press, New York.

Wickham-Crowley, Timothy P.
1992 *Guerrillas and Revolution in Latin America: A Comparative Study of Insurgents and Regimes since 1956*. Princeton University Press, Princeton, New Jersey.

Williams, David
2005 *A People's History of the Civil War: Struggles for the Meaning of Freedom*. New Press, New York.
2010 Civil War Dissent. *New Georgia Encyclopedia*. http://www.georgiaencyclopedia.org/articles/history-archaeology/civil-war-dissent, accessed May 15, 2017.

Williams, Leon
1865 Letter from Leon Williams to Petronia McElroy. Photocopy, Arkansas Archeological Society, Southern Arkansas University Research Station, Magnolia, Arkansas.

Williams, R. H.
1982 *With the Border Ruffians*. University of Nebraska Press, Lincoln.

Wilson, Daniel H.
2001 General Francis Marion: The Swamp Fox and Marine Corps Warfighting Doctrine. Master's thesis, Marine Corps Command and Staff College, Quantico, Virginia.

Winn, Richard
1812 "General Richard Winn's Notes—1780." Peter Force Papers, Series 7E, Reel 3 of 56. Library of Congress, Washington, DC.

Wood, Douglas
ca. 2014 The Shawnee and the Fort Randolph Campaign of 1778. Manuscript on file, Museum of Anthropology, University of Kentucky, Lexington.

Woodrell, Daniel
1998 *Woe to Live On*. Pocket Books, New York.

Worcester, Donald E.
1979 *The Apaches, Eagles of the Southwest*. University of Oklahoma Press, Norman.

Worman, Charles G.
2005 *Gunsmoke and Saddle Leather: Firearms in the Nineteenth Century American West*. University of New Mexico Press, Albuquerque.

Wright, John W.
1926 The Corps of Light Infantry in the Continental Army. *American Historical Review* 31(3):454–461.
1931 "Some Notes on the Continental Army." *William and Mary Quarterly Historical Magazine* 11(3):185–209.

Yaeger, Jason, and Marcello A. Canuto
2000 Introducing an Archaeology of Communities. In *The Archaeology of Communities: A New World Perspective*, edited by Marcello A. Canuto and Jason Yaeger, pp. 1–15. Routledge, London.

Zeisberger, David
1778 Letter to Col. George Morgan, June 9. Draper Manuscripts 3NN120–121. State Historical Society of Wisconsin, Madison. (Reprinted in Kellogg 1916:82–83.)
1999 *David Zeisberger's History of the Northern American Indians*. Edited by Archer Butler Hubert and William Nathaniel Schwarze. Wennawoods Publishing, Lewisburg, Pennsylvania. Manuscript witten 1779–1780. Originally published 1910, *Ohio Archaeological and Historical Quarterly* 19(1/2):1–173.

Contributors

Wade P. Catts is a historian and historical archaeologist with research interests in farmsteads and agricultural landscapes, urban development, conflict archaeology, and Mid-Atlantic regional history and historic preservation.

Carl G. Drexler is an archaeologist with the Arkansas Archeological Survey, specializing in conflict archaeology, historical archaeology, and spatial analysis within archaeology. He has worked for the United States Army, National Park Service, and Colonial Williamsburg Foundation, conducting research in the Southeast, Southwest, Mid-Atlantic, Midwest, and Caribbean.

Clarence R. Geier is a professor of anthropology emeritus at James Madison University. He is the senior editor of four books on the historical archaeology of the Civil War as well as *Historical Archaeology of Military Sites*.

Charles M. Haecker is the staff archaeologist for the National Park Service Intermountain Region's Heritage Partnerships Program in Santa Fe, New Mexico. He specializes in Spanish colonial and military site archaeology in the American Southwest. He has investigated battle sites and encampments associated with the 1540–1542 Vázquez de Coronado expedition, US-Mexican War, Civil War, and Apache and Plains Indian Wars.

Adrian Mandzy is associate professor of history at Morehead State University in Morehead, Kentucky. He has conducted battlefield studies for more than fifteen years and has worked on military sites in Ukraine and the United States. His body of work includes studies of the Battle of Zboriv (1649), the Siege of Zbarazh (1649), Poltava (1709), Blue Licks (1782), the Crater at Petersburg (1864), Lysonia (1916), and Kosmach (1945).

Kim A. McBride works at the Kentucky Archaeological Survey at the University of Kentucky. She has conducted archaeological research on a variety of eighteenth- and nineteenth-century sites, primarily in West Virginia and Kentucky.

W. Stephen McBride is director of interpretation and archaeology at Camp Nelson Civil War Heritage Park in Jassamine County, Kentucky. He has conducted excavations at frontier forts and settlement patterns in the Ohio Valley and Mid-Atlantic and published widely on these studies.

Michael C. Scoggins was the staff historian for the Culture & Heritage Museums (CHM) of York County, South Carolina, and the director of the CHM's Southern Revolutionary War Institute. He is the author or coauthor of eleven books on local and regional history, including *The Day It Rained Militia: Huck's Defeat and the Revolution in the South Carolina Backcountry, May–July 1780*; *Relentless Fury: The Revolutionary War in the Southern Piedmont*; and *Historic York County: An Illustrated History*.

Douglas D. Scott is retired from the National Park Service. He is a visiting research scientist at Colorado Mesa University in Grand Junction as well as an affiliate of the Department of Anthropology at the University of Nebraska–Lincoln. He is particularly noted for his expertise in battlefield archaeology, having worked on more than 50 battlefield sites. He was awarded the Department of the Interior's Distinguished Service Award in 2002 for his innovative research in battlefield archaeology. In 2015 he was the recipient of the J. C. Harrington Medal from the Society for Historical Archaeology for his lifetime achievements in the field of historical archaeology and for his work as a founding leader in the field of conflict and battlefield archaeology.

Michelle Sivilich has more than 15 years' experience in a variety of archaeological settings ranging from seventeenth-century Maryland's first capital, St. Mary's City, to Revolutionary War sites throughout the northeastern United States, to Thomas Jefferson's Monticello.

Steven D. Smith is the director of the South Carolina Institute of Archaeology and Anthropology at the University of South Carolina. He is the coeditor of *Archaeology, History, and Predictive Modeling: Research at Fort Polk, 1972–2002* and *The Southern Colonial Backcountry: Interdisciplinary Perspectives on Frontier Communities*.

Index

Page numbers in italics refer to illustrations.